Anti-Jam GPS Receiver:
Survey on
Architecture, Software Receiver, RF Interference Effects and Mitigation and Signal Acquisition

Mainak Mukhopadhyay

Content

Outline of the Book

Chapter 2 gives an overview of the GPS and its error sources. GPS measurements and their signal structure are explained in brief. C/A-code generation and its auto correlation and cross-correlation properties are explained.

Chapter 3 discusses the GPS receiver architecture. It presents an overview of the work done on a software receiver. It also discusses various acquisition schemes and acquisition detection methods. Research done on weak signal acquisition is presented and the various acquisition performance parameters are also discussed.

Chapter 4 gives an overview of RFI and its effects. It discusses various interference signals and GPS jamming methods. RFI mitigation strategies are discussed briefly.

Chapter 5 discusses the basic acquisition scheme using GPS simulator data and real GPS data.

CHAPTER 1: INTRODUCTION

1.1 Background

The Global Positioning System (GPS) has become a critical part of the navigation infrastructure not only within the United States but also in other nations around the world [Paddan et al., 2003]. Traditionally the GPS was designed for applications where the satellite visibility was not an issue. These GPS receivers were required to have an acquisition sensitivity (minimum signal strength detectable) of -130 dBm [ICD, 2003]. With the E-911 mandate from the Federal Communications Commissions (FCC), it has become necessary to provide positions under all kinds of environments [FCC, 2003]. Indoor and urban canyon environments typically attenuate the GPS signal by about 20-25 dB [MacGougan, 2003]. Thus a signal strength of -150 dBm should be able to be acquired and tracked by a GPS receiver to provide position. A GPS signal below the -135 dBm power level is categorized as a weak signal [Tsui and Lin, 2001]. GPS receivers designed to operate at nominal GPS signal strengths are referred to as standard GPS receivers, while the receivers designed for weak signal environments are called the high-sensitivity receivers [Tsui and Bao, 2000].

A GPS receiver must detect the presence of the GPS signal to track and decode the information from the GPS signal required for position computation [Kaplan, 1996]. Tracking of the signals is possible only after they have been acquired, so acquisition is the first step in the GPS signal processing scheme. The acquisition process must ensure that the signal is acquired at the correct code phase and carrier frequency [Spilker and Parkinson, 1996]. A GPS receiver should be capable of giving a reasonably correct position (within 5-10 m) in the presence of interference and multipath signals. Thus, the GPS receiver should be capable of mitigating the effects of Radio Frequency (RF) interference and multipath signals [Maenpa et al., 1997].

Any radio navigation system can be disrupted by an interference of high power and GPS is no exception. Although the GPS frequency bands are protected by the FCC frequency assignments, there is still a chance of spurious unintentional and intentional interference [Spilker and Parkinson, 1996]. RF interference (RFI) is a major source for the degradation of GPS accuracy and reliability. The interference signals must be mitigated to prevent the GPS receiver from giving erroneous information. This becomes important when the GPS is being used for critical applications [RTCA, 2001]. RFI mitigation can be done at various stages in the GPS receiver. Interference signals can be filtered out either by the GPS antenna or the front-end section of the GPS receiver [Littlepage, 1999]. The GPS signal processing and navigation algorithms can be modified to estimate the interference effect and detect the interference source [Macabiau et al., 2001].

1.2   Relevant Research

GPS signal acquisition has been extensively studied since the launch of the GPS program. A simple time domain correlation approach was widely used in the first generation GPS receivers [Spilker and Parkinson, 1996]. The time domain correlation methods were sequential in nature and simple for hardware implementation. The implementation aspects of the signal acquisition schemes using Field Programmable Gate Arrays (FPGA) have been extensively studied by Gunawardena [2000], Alaqeeli and Starzyk [2001] and Alaqeeli [2002]. The GPS receiver manufacturers have used different technologies and modifications of this scheme in their receivers and most of these methods are the intellectual property (IP) of the respective companies. Van Nee and Conen [1991] pioneered the study of GPS signal acquisition in the frequency domain whereby they developed a circular convolution technique to speed up the acquisition process. Tsui and Bao [2000] improved the scheme to use only half the GPS signal spectrum for acquisition. Frequency domain methods allow the correlator section of the GPS receiver to be implemented in software. Software receiver design and development was studied by Akos et al. [2001], Burns et al. [2002] and Ledvina et al. [2003].

The use of the GPS in weak signal environments such as urban canyons, forest areas and indoors developed a need to acquire the GPS signals about 20-25 dB below the nominal signal strength. Chansarkar [2000], Choi et al. [2002] and Lin et al. [2002] developed techniques to extend the signal integration period during acquisition beyond the navigation data bit duration to increase the acquisition sensitivity. Tsui and Lin [2001] developed a scheme to have different thresholds for signal detection under different environments. Weak signal acquisition is required to provide a Doppler estimate within the bandwidth of the tracking loops. Tsui and Bao [2000] and Akopian et al. [2002] developed fine frequency estimation methods to acquire the signals with a resolution up to 1 Hz.

GPS is rapidly becoming the most widely used navigation system in automobile navigation, personal navigation, defence applications, timing applications and atmospheric studies. Methods of improving its accuracy through the use of Differential GPS (DGPS) has further opened up applications in precision navigation, including air, sea, and land. These applications require GPS to provide a reliable and accurate solution, however the system is vulnerable to low power interference from RF signals in the GPS frequency band [Littlepage, 1999]. Unintentional interference and jamming are two of the major concerns in using the GPS for various critical applications [Kaplan, 1996]. The Federal Aviation Authority (FAA) sponsored various tests to determine the vulnerability of GPS receivers to RFI allowing it to establish the interference tolerance standards for GPS receivers in civil aviation. These tests were focused on the coarse/acquisition (C/A) code receiver tracking degradation and loss of lock under different interference conditions. RFI effects on GPS signals have been extensively studied by researchers since the time of designing the GPS system (1970s).

Johannessen et al. [1990] studied potential sources of interference for the GPS and provided some solutions for civil aviation applications. Littlepage [1999] analyzed the effect of various interference signals on the use of the GPS for civil applications. RTCA [2001] established the minimum operational performance standards (MOPS) for the GPS/WAAS receivers under interference conditions based on the results from the FAA tests. Erlandson and Fraizer [2002] studied the effect of RFI signals on GPS in marine applications. Buck and Sellick [1997] analyzed the interference caused by television (TV) signals.

Different RFI mitigation techniques have been developed over the years by different researchers. These techniques can be classified into five categories

1. Antenna gain variation

2. RF filtering

3. Interference location

4. Sampling and Automatic Gain Control (AGC)

5. RFI mitigation in tracking

Antenna gain can be varied to provide a zero gain in the direction of the interference signal. This ensures that the interference signal is not captured by the GPS antenna. Different adaptive antenna arrays have been developed to achieve this goal. Bond and Brading [2000] developed a direction finder (DF) location vector to null the gain for interference signals. Kunysz [2001] developed a controlled rejection pattern antenna (CRPA) array to provide better tolerance to various kinds of interference signals. Navsys Inc. first developed a commercial GPS antenna to include spatial signal processing. Brown et al. [2000] further enhanced the antenna to detect the presence of interference signals and to estimate its direction. Moore and Gupta [2001] analyzed antenna arrays equipped with a space-time adaptive processor (STAP) to provide interference mitigation. The drawbacks in a STAP antenna were overcome using space-frequency adaptive processing (SFAP) [Gupta and Moore, 2001]. Vaccaro and Fante [2000] studied the different adaptive processing algorithms under an interference environment to determine the best methods available for the GPS antenna design.

RF filtering can be used to filter out interference signals outside the GPS frequency band. These filters should have a sharp cut-off outside the GPS bandwidth, low loss in the pass band and high rejection in the stop band. Escobar and Harper [2001] designed some high temperature superconducting (HTS) filters to provide the interference mitigation for tactical mobile applications.

Locating and nullifying the source(s) of the interference can realize the mitigation of the errors. Various interference localization techniques have been developed for determining the source(s) of the interference. Gormov et al. [2000] developed an inverse long range radio navigation (LORAN) method to estimate the direction of the jammer location. Brown et al. [1999] developed a time difference of arrival (TDOA) method to determine location of a large number of jammer sources. Shau-Shinu and Enge [2001] developed the RFI location method using a network of distributed sensors. The advantage of using a network of distributed sensors is that no sensor motion is required and is robust to sensor failures.

Bastide et al. [2003] studied the AGC to use it as a tool for interference assessment. An AGC is an accurate indicator of the noise in the receiver and variations of its threshold levels can be used to determine the presence of an interference signal. Blanking of the GPS signal using the AGC can be used to eliminate pulse interference. Hegarty et al. [2000] studied the effect of pulse interference on the AGC and designed a technique to suppress the pulse signal and determine the loss of signal during blanking. Leica Inc.

developed an RFI mitigation technique using multi-level sampling. Maenpa et al. [1997] analyzed the technique and found it suitable to minimize in-band interference.

Macabiau et al. [2001] devised a multi-correlator technique for detecting continuous wave (CW) interference. Manz et al. [2000] developed a technique to mitigate interference in the phase lock loop (PLL) provided the user is stationary and has a stable clock. Cooper and Daly [1997] developed a technique of preprocessing the GPS signal to remove the interference components using a PLL before passing the signal to the GPS tracking loops. These techniques were effective in mitigating CW interference but are not suitable for different interference signals.

A software receiver allows flexibility in dealing with interference. The exploitation of spectrum transforms as well as other mathematical tools are more feasible in software than in traditional hardware receivers. The accuracy of this representation is a function of the signal bandwidth, sampling rate and quantization error. Cutright et al. [2003] developed a frequency domain approach using a software receiver to mitigate RFI. Burns et al. [2002] evaluated software receiver interference mitigation by varying the number of bits in an Analog-to-Digital converter (ADC).

A considerable amount of research has been done on the GPS signal acquisition process and various RFI mitigation techniques. GPS signal acquisition has been studied for feasibility in hardware and software implementation, weak signal environments and fine frequency estimation. RFI mitigation at various stages in a GPS receiver from the GPS antenna to the navigation solution has been studied extensively except for the acquisition process. RFI mitigation in the acquisition process has not been the focus of study mainly because of the few parameters controlling the acquisition process. This research studies the effect of the RFI signals in the acquisition process of a GPS receiver.

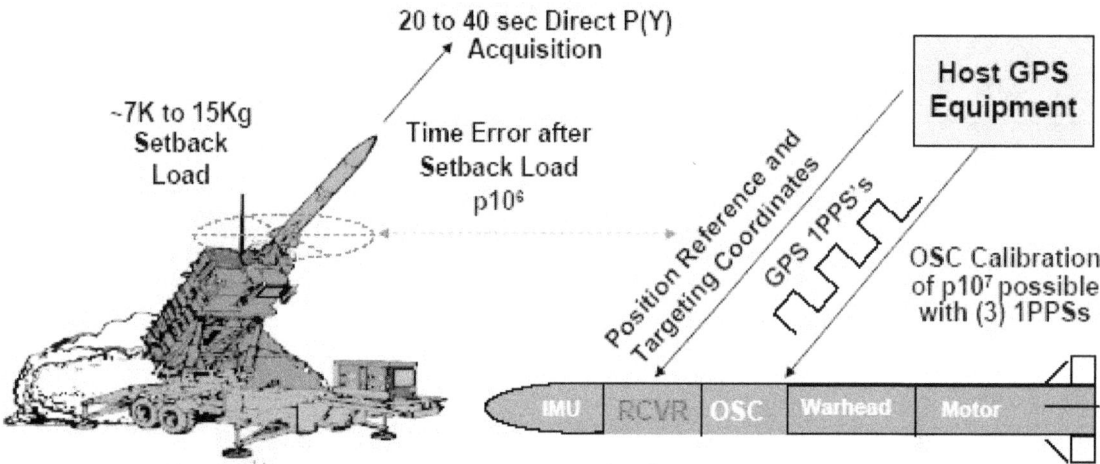

Weapons Platforms

1 to 10 ms Time Error for Initialization of the Direct P(Y) Receiver

Typical Initialization Parameters for Direct P(Y) Acquisition

Position Uncertainty	a few meters SEP
Velocity Uncertainty	N/A
Time Uncertainty	± 10s of microseconds
Almanac/Ephemeris	Most recent from Host

Technology	Max Gain[1]	Number of Emitters	Estimated Cost	Remarks
IMU Receiver Code Loop Aiding	10 dB	N/A	$10 – 40 K	Cost depends on accuracy; higher cost represents 1 nmi/hr quality
Adaptive Controlled Radiation Pattern Antennas (CRPA)	35 dB	~(# elements –1), but Depends on geometry	$2 – 20 K	Less capable systems available now; higher end systems not in production for a few years
Low-Elevation Antenna Nuller (LEAN)	35 dB	Any Number Near Horizon	$3 K	Still in development; need to assess impact on satellite tracking
Signal Polarization Cancellation Antenna	31 dB	14 dB for 4 Broadband	$3 – 5 K	L1 C/A available
Reference Canceller	50 dB	Any Number Near Horizon	-	In development; need to assess impact on satellite tracking
Adaptive Filtering or Narrowband Frequency Excision (FX)	50 dB	3-20 Narrowband	<$100	Ineffective against broadband interference
Combined FX & Nonlinear Adaptive Processing (FXNONAP)	40 dB	20 Narrowband, up to 3 Broadband	<$100	NONAP deployed in sub fleet; FXNONAP still in development
Direct Measurement Processing	20 dB	N/A	-	In development

[1] Actual performance highly dependent on scenario

CHAPTER 2: GPS SYSTEM OVERVIEW

This chapter gives a brief overview of the GPS system, various GPS measurements and their signal structure. It also discusses C/A-code generation and its auto correlation and cross correlation properties.

2.1 GPS System

The GPS is a satellite-based positioning system capable of providing a user position anywhere in the world. This system was developed by the Department of Defense (DoD) to support the military forces of the United States of America by providing world-wide, real-time positions [Parkinson et al., 1995]. GPS can be used for civilian applications even though it was developed for military applications [Spilker and Parkinson, 1996]. The system currently consists of 27 (nominally 24) satellites which provide continuous information for the user to compute position, velocity and time (PVT). The satellites orbit about 28,000 km above the Earth's surface and have an orbital period of 11 hr 58 m [ICD, 2003]. The GPS functions on the concept of one-way Time-of-Arrival (TOA) ranging whereby the user determines the TOA of the GPS signal transmitted by the GPS satellites. These ranges are used by the user to compute the navigation solution. A 3D position computation requires the range information from at least three satellites [Kaplan, 1996]. However, the GPS receiver clock is not generally synchronized with the satellite clocks and hence an additional measurement is required to solve the receiver clock offset.

GPS provides different accuracy levels for civilian and military users. Civilian users have access to the C/A-code which provides the Standard Positioning Service (SPS). Military users use a Precise (P)-code to get the Precise Positioning Service (PPS). The P-code is encrypted and hence not available for civilian users. The SPS provides an accuracy of 36 m (2D RMS 95%) in the horizontal plane and 77 m (95%) in the vertical direction [Stenbit, 2001] although recent field tests show accuracies of 5-10 m (1-σ RMS) [MacGougan, 2003]. GPS operates on two signal frequencies using code division multiple access (CDMA) technology to transmit the ranging codes [ICD, 2003]. The GPS signal structure is discussed in Section 2.3.

2.2 GPS Observations and Error Sources

Three different types of positioning information can be extracted from a GPS satellite signal, namely a pseudorange measurement, a carrier phase measurement, and the Doppler.

2.2.1 Pseudorange

A pseudorange is a range measurement between the GPS satellite and the user. This range measurement has inherent errors which make it different from the true range [Kaplan, 1996]. The pseudorange is a measure of the time delay required to align the GPS signal received from the satellite with the local GPS signal generated by the receiver. This time delay is converted into a distance measurement using the speed of light. The receiver clock and satellite clock are not synchronized which introduces an error in the range. Hence the measured range is different from the true range and is called a pseudorange [Spilker and Parkinson, 1996]. The pseudorange is instantly available to compute position information and is given by Equation (2.1) [Wells et al., 1986].

$$p(t) = \rho(t) + d_{orb} + c\big(dt(t) - dT(t)\big) + d_{trop}(t) + d_{iono}(t) + \varepsilon_p \qquad 2.1$$

where

$p(t)$ is the pseudorange measurement at time t (m),

$\rho(t)$ is the true distance between satellite and receiver (m),

d_{orb} is the orbital error (m),

c is the speed of light (m/s),

$dt(t)$ is the satellite clock error (s),

$dT(t)$ is the receiver clock error (s),

$d_{trop}(t)$ is the tropospheric error (m),

$d_{iono}(t)$ is the ionospheric error (m), and

ε_p is the code multipath and measurement noise (m).

2.2.2   Carrier Phase

A carrier phase measurement is a range measurement computed from the GPS carrier signal information. The total number of the carrier cycles from the GPS satellites to the user are measured and converted into a range measurement using the carrier wavelength [Kaplan, 1996]. The receiver cannot determine the number of integer cycles before the signal is acquired. This is referred to as the integer cycle ambiguity. This ambiguity must be resolved before the carrier phase measurement can be used for position computation. It can be represented by Equation (2.2) [Wells et al., 1986].

$$\theta(t) = -\lambda\varphi(t) = \rho(t) + d_{orb} + c(dt(t) - dT(t)) + d_{trop}(t) - d_{iono}(t) + \lambda N + \varepsilon_\theta \qquad 2.2$$

where

$\theta(t)$ is the carrier phase measurement at time t (m),

$\varphi(t)$ is the carrier phase measurement (cycles),

λ is the carrier wavelength (m/cycle),

N is the integer carrier phase ambiguity (cycles), and

ε_θ is the carrier multipath and measurement noise (m).

The definitions of the other symbols are the same as in Equation (2.1). The carrier phase measurement with the ambiguity resolved to the correct integer provides a very accurate range measurement and is used to provide centimetre-level position accuracies.

2.2.3 Doppler

The Doppler is a measure of the instantaneous rate of the GPS carrier phase and is the instantaneous Doppler frequency shift of the incoming carrier. The Doppler shift results from the relative motion between the receiver and the satellite. The major role of the Doppler measurement in the navigation process is to compute a velocity estimate.

2.2.4 GPS Errors

GPS measurements have various errors including satellite clock errors, orbital errors, atmospheric errors, receiver clock error, multipath and interference [Wells et al., 1986]. The satellite clock error is the drift in the satellite clock with respect to the GPS time reference. The GPS master control station synchronizes the satellite clock with the GPS clock during the upload of the navigation information, and this offset is transmitted in the navigation message. The satellite orbital error is the difference between the satellite's position using the ephemeris and the actual values [ICD, 2003].

When the GPS signal travels through the troposphere its path will bend slightly due to the refractivity of the troposphere [Kaplan, 1996]. The change of the refractivity from free space to the troposphere causes the speed of the GPS signal to slow down which results in a delay of the GPS signal. This tropospheric delay is a function of the temperature, pressure, and relative humidity [Spilker and Parkinson, 1996]. Hopfield [1969] and Saastamoinen [1972] have developed different tropospheric delay models which can reduce the tropospheric error by about 90%.

The ionosphere is the layer of the atmosphere that extends from 60 to over 1000 km of height above the Earth's surface. It is an important source of range and range-rate errors for GPS users requiring high-accuracy measurements [Tsui and Bao, 2000]. The ionospheric variation is generally large compared to the troposphere and is more difficult to model. Ionospheric error can be eliminated using dual frequency measurements from

GPS. The single frequency ionospheric (Klobuchar) model described in ICD [2003] can reduce the ionospheric error by 50%. Ionospheric error can be further reduced using better ionospheric estimation models and Wide Area Augmentation Systems (WAAS) can be used to provide ionospheric corrections to reduce the error [Lachapelle, 2002].

The user clock is often inaccurate and not synchronized with the GPS clock, which results in the user clock error. The approximate magnitudes of the different errors are listed in Table 2.1.

Table 2.1: GPS error sources [Lachapelle, 2002]

GPS error source	Error magnitude (1 σ) (m)
Satellite clock and orbital errors	2.3
Ionosphere on L1	7.0
Troposphere	0.2
Code multipath	0.01-10
Code noise	0.6
Carrier multipath	50×10^{-3}
Carrier noise	$0.2-2 \times 10^{-3}$

All errors except multipath and noise can be reduced using techniques such as single-differencing, double-differencing and DGPS corrections [Lachapelle, 2002]. Multipath is the error caused by the reflected GPS signals entering the receiver front-end and mixing with the direct signal [Braasch and Van Grass, 1991]. Its effect will be more pronounced for static receivers close to large reflectors. It is specific to a receiver/antenna and depends on the surrounding environment. Hence care has to be taken while installing GPS receivers for static applications, such as reference stations.

2.3   GPS Signal Structure Overview

The current GPS signal structure was developed specifically for positioning purpose and since it was developed for military application it also required a good resistance to jamming signals [Parkinson et al., 1995]. The spread spectrum concept was used to transmit ranging codes to provide the desired anti-jamming performance. A pseudo random noise (PRN) sequence with a high chipping rate was used to transmit the navigation information on to the GPS frequencies [ICD, 2003]. Spread spectrum signals have power below the noise level and can be recovered only with an appropriate spreading code. The two spreading codes used in the GPS signal are the C/A-code and P-code. These spreading codes were selected from a family of Gold codes [Kaplan, 1996]. Each satellite transmits the signal on the two frequencies (L1 and L2) with the P-code present on both the frequencies. The C/A-code is transmitted only on the L1 frequency. The CDMA technique of transmitting different spreading code for each satellite on the same frequency is used in the GPS to distinguish the signals from the different satellites [ICD, 2003]. Figure 2.1 represents the current GPS signal structure. The basics of spread spectrum and CDMA are discussed briefly in Sections 2.3.1 and 2.3.2.

17

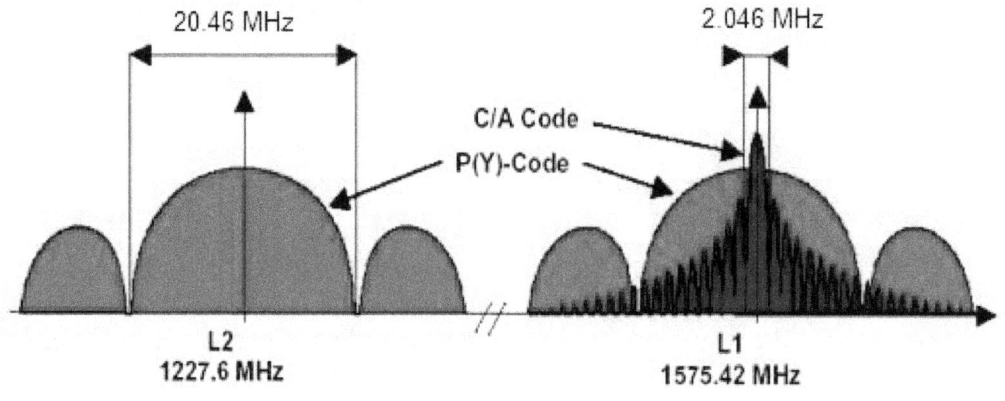

Figure 2.1: GPS signal spectrum

2.3.1   Spread Spectrum Basics

The spread spectrum concept consists of transmitting the information over a large bandwidth and using a PRN sequence to spread the information [Peterson et al., 1995]. The amount of bandwidth required for transmission is determined by the PRN sequence bandwidth. All modulation techniques which use a bandwidth wider than required for transmission are not spread spectrum techniques. The spread spectrum technique is useful for long distance communication with less interference problems [Kaplan, 1996]. During recovery of the spread spectrum signal, any interference signal is spread thereby reducing its power level below the noise. Spread spectrum solves two important communication problems namely pulse jamming and low probability of detection [Peterson et al., 1995]. The pulse jammer power level is reduced during signal recovery in the spread spectrum method. The spread spectrum can be recovered only when the PRN signal used for spreading is known [Peterson et al., 1995]. This reduces the chance of signal detection by other users in the same frequency band.

A direct sequence (DS) spread spectrum is used for the GPS signals. It consists of modulating the information signal using a spreading carrier signal [Peterson et al., 1995]. A binary phase shift keying (BPSK) signal is used to spread (modulate) the navigation data signal. The BPSK signal is a square wave (±1) and the phase of the modulated signal changes by 180 degrees with a change in the sign of the signal. Consider a data modulated carrier signal, S(t), given in Equation (2.3).

$$S(t) = A\cos(\omega t + \Phi(t)) \qquad\qquad 2.3$$

where

        A   is the amplitude of the carrier signal (volts),

        ω is the carrier frequency (Hz), and

        Φ is the data modulation signal.

BPSK spreading is performed by multiplying the S(t) by a function c(t), which represents the spreading waveform and the resulting signal, $S_t(t)$, is given in Equation (2.4).

$$S_t(t) = Ac(t)\cos(\omega t + \Phi(t)) \qquad\qquad 2.4$$

This spread spectrum signal is then transmitted and is received by the receiver after a delay of T. To recover the signal, the receiver must replicate the spreading signal used at the transmitter and match the phase of the spreading signal. The received signal, $S_r(t)$, is given in Equation (2.5).

$$S_r(t) = Ac(t-T)\cos(\omega t + \Phi(t-T) + \phi) \qquad\qquad 2.5$$

where

        ϕ is the random phase error (radians).

The spreading signal, c(t), has values of ±1, which when multiplied with the received signal c(t-T) will have a value of one when the phase of the replica signal matches the incoming signal. This allows for the recovery of the information in Equation (2.5) except for some random phase error [Tsui and Bao, 2000]. A similar concept is used in the GPS for transmission and recovery of the information.

2.3.2   Code Division Multiple Access

A CDMA system is one in which different transmitters transmit the information on the same carrier frequency using different spreading codes to distinguish each transmitter. The spreading codes used are a set of orthogonal or near-orthogonal codes [Kaplan, 1996]. An orthogonal code has a zero correlation with the other codes used in the system. The GPS uses the CDMA technology to transmit information from the GPS satellites at the same centre frequency which gives rise to the possibility of interference among the codes [Tsui and Bao, 2000]. The codes do not have zero cross-correlation due to side lobes of the codes and hence there is a possibility of a cross-correlation peak, resulting from correlation between same or different codes, being higher than the autocorrelation peak when the desired signal is weak.

2.3.3   GPS Signal Structure

GPS satellites transmit on two frequencies in the L-band of the frequency spectrum called L1 and L2 signals. The L1 signal is the primary frequency and is transmitted at 1.57542 GHz and L2 is the secondary frequency and is transmitted at 1.2276 GHz. The GPS signal is a BPSK DS spread spectrum signal [ICD, 2003]. The two carrier frequencies are modulated by the spread spectrum codes with a unique PRN associated with each space vehicle (SV). The signals are further modulated by a 50 Hz navigation data message [ICD, 2003]. The C/A and P-codes are in phase quadrature with each other on the L1 frequency. A C/A-code is 1023 bits long and is available to civilian users. A P-code is one week long code and the structure of the P-code is known. It is reserved for military applications and hence is encrypted using a Y-code. This encrypted code is transmitted instead of the P-code on both frequencies [ICD, 2003].

Figure 2.2 shows a block diagram of the GPS satellite transmitter unit. The GPS satellite uses a 10.23 MHz reference clock to generate both the L1 and L2 frequencies. This clock is usually a cesium clock and generates a clock frequency slightly lower than 10.23 MHz to account for the relativistic effect [Spilker and Parkinson, 1996]. The GPS signal broadcast on the L1 and L2 frequencies have the signal structure given in Equations (2.6) and (2.7) [Kaplan, 1996].

$$L_1(t) = A_1 P(t) N(t) \cos(2\pi f_1 t) + A_1 C/A(t) N(t) \sin(2\pi f_1 t) \qquad 2.6$$

$$L_2(t) = A_2 P(t) N(t) \cos(2\pi f_2 t) \qquad 2.7$$

where

A_1 is the L1 signal amplitude,

A_2 is the L2 signal amplitude,

$P(t)$ is the P-code,

$C/A(t)$ is the C/A-code,

$N(t)$ is the navigation data,

$\cos(2\pi f_1 t), \cos(2\pi f_2 t), \sin(2\pi f_1 t)$ are the unmodulated L_1 and L_2 signals, and

$L_1(t)$ and $L_2(t)$ are the modulated L_1 and L_2 signals.

21

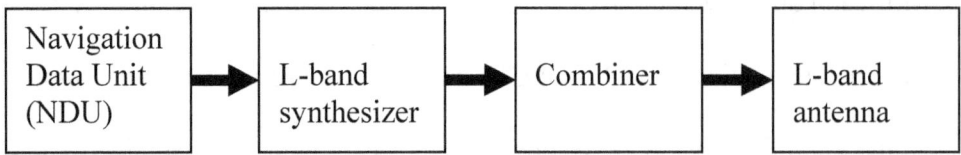

Figure 2.2: GPS satellite transmitter unit [Spilker and Parkinson, 1996]

The navigation data unit (NDU) generates the cosine and sine of the carrier signal which are modulated by a 50 Hz navigation data signal. This modulated signal is then spread using the C/A-code and the P(Y)-code [Kaplan, 1996]. The NDU block performs the function of modulating the signal, and the synthesizer is used to manipulate the signals according to the bandwidth specifications of the signal. For the L1 signal, the combiner combines the C/A-code and the P(Y)-code signals onto one signal. Both the L1 and the L2 signals are transmitted to the Earth using an L-band antenna.

2.4 C/A Code Generation

A block diagram of the C/A-code generator is shown in Figure 2.3. The C/A-code is generated using a linear code generator. Linear code generators can be described by a polynomial of the form $1 + \sum_{i=0}^{n} X^i$, where X^i means that the output of the ith cell of the n-stage shift register is used as the input to a modulo-2 adder and the 1 means that the output of the adder is fed to the first cell [Tsui and Bao, 2000]. The C/A-code generator consists of two 10-bit shift registers (G1 and G2), which generate a maximum length pseudo noise (PN) codes with length of $2^{10} - 1 = 1023$ bits. The only state the shift register must not get into is an all-zero state. The shift registers can be described by the polynomials $G1 = 1 + X^3 + X^{10}$ and $G2 = 1 + X^2 + X^3 + X^6 + X^8 + X^9 + X^{10}$ [ICD, 2003].

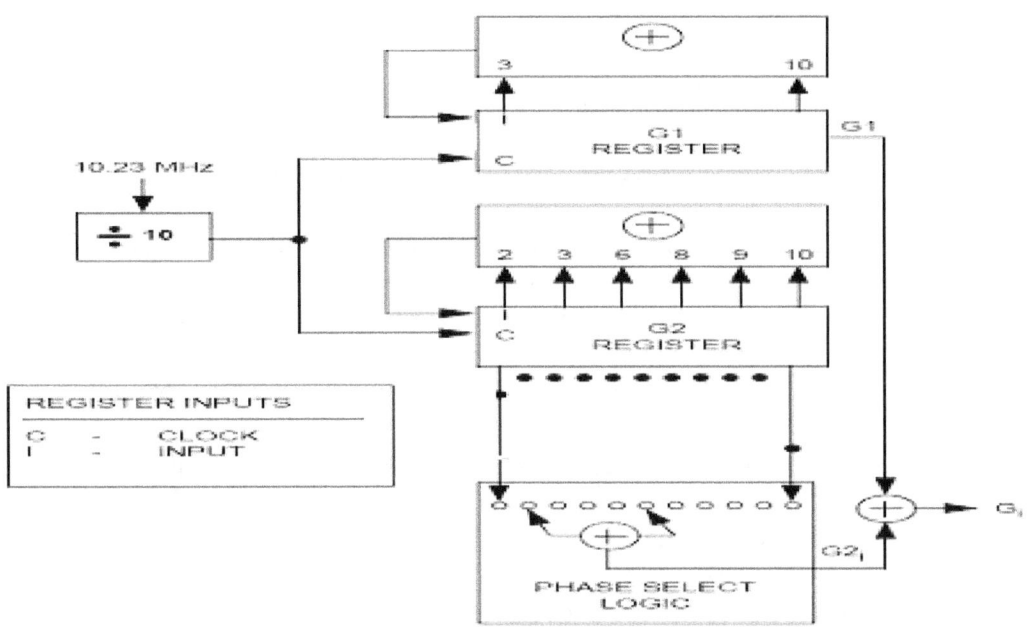

Figure 2.3: GPS C/A code generator [ICD, 2003]

The unique C/A-code for each SV is a result of the exclusive-or of a delayed version of the G2 output sequence and the G1 direct output sequence. The delay effect in the G2 PRN code is obtained by the exclusive-or of the selected positions of two taps whose output is called G21. This is because a PN code sequence has the property that when added to a phase-shifted version of the same code it does not change but obtains another phase [Tsui and Bao, 2000]. The function of the two taps on the G2 shift register is to shift the code phase in G2 with respect to the code phase in G1 without the need for an additional shift register to perform this delay. Each PRN code is associated with two tap positions on the G2 register. Table 2.2 describes these tap positions for all defined GPS PRN numbers and also specifies the equivalent delay in the C/A-code chips [ICD, 2003]. The chipping rate for the C/A-code is 1.023 MHz and hence the C/A-code repeats every millisecond.

The first 32 of these PRN numbers are reserved for the GPS satellites and the remaining PRNs (33 to 37) are reserved for other uses such as ground transmissions. The generation of the P-code is more complex than the C/A-code. P-code generators use four 12-bit shift registers and their sequences are combined to generate the P-code. The sequence generated is 38 weeks long which is partitioned into 37 unique sequences that are truncated at the end of one week. Each week long code represents the P-code for the GPS satellites [ICD, 2003]. The P-code is not a part of this research and hence the generation of the P-code is not discussed in detail.

Table 2.2: GPS C/A-code delay [ICD, 2003]

PRN number	C/A code tap selection	C/A code delay (chips)	PRN number	C/A code tap selection	C/A code delay (chips)
1	2Θ6	5	20	4Θ7	472
2	3Θ7	6	21	5Θ8	473
3	4Θ8	7	22	6Θ9	474
4	5Θ9	8	23	1Θ3	509
5	1Θ9	17	24	4Θ6	512
6	2Θ10	18	25	5Θ7	513
7	1Θ8	139	26	6Θ8	514
8	2Θ9	140	27	7Θ9	515
9	3Θ10	141	28	8Θ10	516
10	2Θ3	251	29	1Θ6	859
11	3Θ4	252	30	2Θ7	860
12	5Θ6	254	31	3Θ8	861
13	6Θ7	255	32	4Θ9	862
14	7Θ8	256	33	5Θ10	863
15	8Θ9	257	34	4Θ10	950
16	9Θ10	258	35	1Θ7	947
17	1Θ4	469	36	2Θ8	948
18	2Θ5	470	37	4Θ10	950
19	3Θ6	471			

Θ   Exclusive-OR operator

2.5   C/A Code Correlation Properties

This section discusses the correlation properties of the C/A-code and their probability of occurrence.

2.5.1   Auto Correlation

The C/A-code correlation properties are fundamental to the signal acquisition and demodulation processes in a GPS receiver [Spilker and Parkinson, 1996]. The correlation of a code with itself is called autocorrelation, while the correlation between two codes is called cross-correlation. The autocorrelation function involves replicating the code and then shifting its phase while multiplying it with the original function. When the phases of the two signals match, the maximum correlation is obtained. The autocorrelation function for a Pseudo Noise (PN) sequence, PN(t), whose amplitude is ±A, chipping period is T_c and period is NT_c is given by Equation (2.8) [Macabiau et al., 2001].

$$R(\tau) = \frac{1}{T_c} \int_0^{T_c} PN(t)PN(t+\tau)dt \qquad\qquad 2.8$$

A PN sequence of length $N = 2^n - 1$, where n is the number of shift register stages used to generate the sequence is called a maximum length sequence [Kaplan, 1996]. The autocorrelation function yields $-A^2/N$ outside the correlation interval because the number of negative values (-1) is always one greater than number of positive values (+1) in a maximum length PN sequence [Peterson et al., 1995]. An autocorrelation function for a maximum length PN sequence is the infinite series of triangular functions with period NT_c. The negative correlation amplitude $(-A^2/N)$ is obtained when the phase shift, τ, is greater than $\pm T_c$, (or multiples of $\pm T_c(N\pm1)$) and represents a dc term in the series [Macabiau et al., 2001].

GPS PRN codes have periodic correlation triangles and a peak spectrum that has similar characteristics to the maximum length PN sequences [Kaplan, 1996]. However the GPS codes are not maximum length PN sequences. A simple 10-bit linear code generator can generate 1023 sequences but all the autocorrelation functions have considerable power in the side lobes which affects the signal detection at low signal strengths. This problem was overcome by combining sequences from two 10-bit shift registers (G1 and G2) to generate the C/A-code [Spilker and Parkinson, 1996]. The combination of two sequences from the C/A-code generator yields 1023 possible combinations. The correlation properties of these sequences were studied and 32 codes with the best cross-correlation properties were selected for the GPS satellites [Kaplan, 1996].

The autocorrelation function of the GPS C/A-code has the same period and shape in the correlation domain as the maximum length PN sequences. However, there are small correlation values in the interval between the maximum correlation intervals. These small fluctuations in the autocorrelation function of the C/A-code result in the deviation of the line spectrum from the $\sin(x)/x$ envelope [Spilker and Parkinson, 1996]. The 1 KHz line spectrum spacing is the same for all the C/A-codes and the 10-bit maximum length sequence code. The ratio of power in each of the C/A-code line spectrum to the total power can fluctuate by nearly 8 dB with respect to the -30 dB levels that would be obtained if every line contained the same power [Kaplan, 1996]. The autocorrelation for PRN 12 is shown in the Figure 2.4.

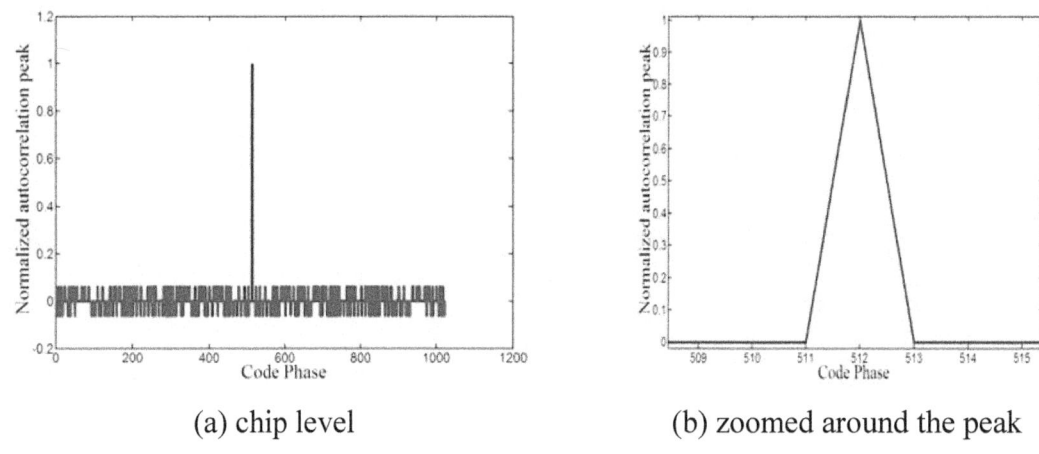

(a) chip level   (b) zoomed around the peak

Figure 2.4: Autocorrelation plot for SV 12

2.5.2   Cross Correlation

A GPS receiver should generate a replica of the GPS PRN code and shift its phase to align with the PRN code for each SV. The PRN codes for different satellites should have poor cross-correlation properties among them to allow acquisition of the correct PRN signal. The GPS C/A-code length is 1023 chips which causes the cross-correlation properties to be poor for some codes. The C/A-code autocorrelation peaks are higher than cross-correlation peaks by just 21-24 dB, which can cause false acquisition [Kaplan, 1996]. Table 2.3 lists the C/A-code cross correlation power probabilities.

Table 2.3: Cross correlation probability of C/A code [Kaplan, 1996]

Cumulative Probability of Occurrence	Cross correlation for any two codes (dB)
0.23	-23.9
0.50	-24.2
0.99	-60.2

The P-code is not a maximum length sequence but since its period is very long its autocorrelation and cross-correlation properties are almost ideal. The cross-correlation peak between the P-codes is 127 dB lower than the autocorrelation peak, which is much better compared to 24 dB difference for the C/A-codes [Kaplan, 1996]. The autocorrelation function of the P-code has similar characteristics to the C/A-code. The study of P-code is not a part of this research and hence the correlation properties of the P-code will not be discussed.

CHAPTER 3: GPS RECEIVER ARCHITECTURE AND SOFTWARE
RECEIVER DESIGN

This chapter discusses the architecture of a conventional GPS receiver and it presents an overview of the work done on a software receiver. Research done on the acquisition process including weak signal acquisition, is presented and various acquisition performance parameters are also discussed.

3.1 Conventional GPS Receiver Architecture

A conventional GPS receiver consists of three blocks which process the incoming GPS signal in three different frequency ranges. The RF section operates on the incoming GPS signals at the GHz frequency range, the signal processing section operates on the signal at the MHz/KHz frequency range and the data processing section operates at the Hz frequency range. A conventional GPS Receiver block diagram is shown in Figure 3.1.

The RF section is responsible for receiving the GPS signal from the antenna and down converting it to an intermediate frequency (IF) [Kaplan, 1996]. The down conversion process can be performed in a single stage or in multiple stages. Each stage consists of a local oscillator, mixer and band pass filter to eliminate the undesired mixer product. The RF section amplifies the signal and also determines its precorrelation bandwidth. The IF signal is sampled at a desired sampling rate using an AGC and an ADC [Tsui and Bao, 2000].

30

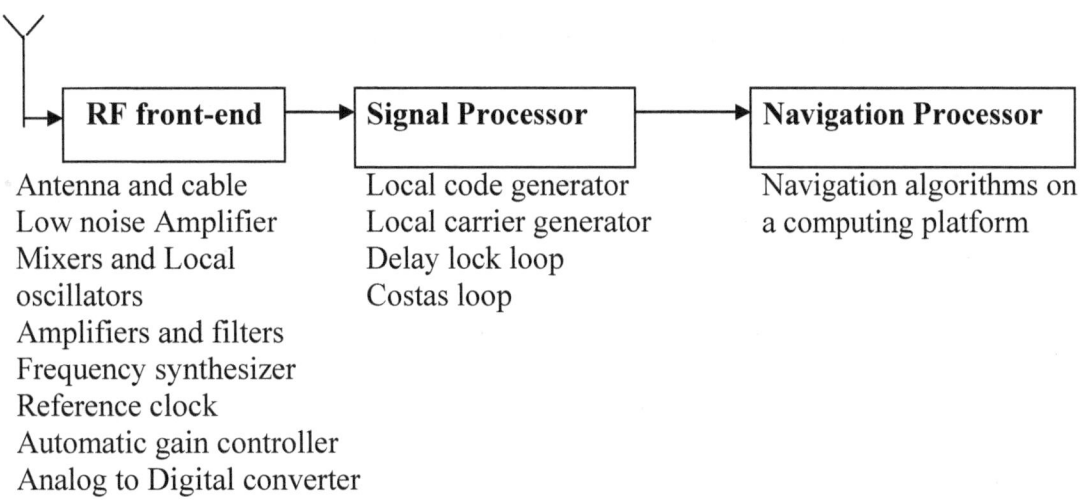

Figure 3.1: GPS receiver architecture

The signal processor acquires and tracks the signals and determines the navigation data bit value. Acquisition involves performing a two-dimensional search in the code and Doppler range. It involves a carrier wipe-off wherein the carrier from the incoming GPS signal is removed and code wipe-off wherein the PRN code from the incoming GPS signal is removed. Once the carrier is wiped off, the residual frequency component is the Doppler. The acquisition process must replicate both the carrier and code of the satellite in order to acquire it (the signal match for success is two-dimensional). To acquire the signal, correlation is done over a period called the predetection integration period, which is chosen depending on the acquisition scheme, time-to-first-fix (TTFF) requirement, data bit prediction and Doppler frequency [Tsui and Bao, 2000]. When the replica signal correctly matches the code and Doppler of the received signal, a GPS signal peak is obtained. This peak is easily distinguishable from other peaks at the nominal power (-130 dBm) and allows signal acquisition. The GPS signal acquisition process is explained in detail in Section 3.3.

Once the signal is acquired, the tracking loops are used to keep lock on the signal and to detect the navigation data bit transitions. A PLL and a Frequency lock loop (FLL) are used to track the carrier signal whereas a Delay lock loop (DLL) is used to track the code phase [Spilker and Parkinson, 1996]. This section generates the pseudorange and the Doppler measurements, computes the Carrier-to-Noise (C/No) ratio of the signal to determine signal quality and determines the thresholds for the acquisition and tracking process. It also extracts the raw navigation data from the data bits collected.

The navigation processor extracts the navigation information from the raw data bits collected, computes satellite positions and uses them to compute the user's PVT information. Present day GPS receivers combine the receiver blocks to reduce cost and size and to have a greater level of integration [Ray, 2003]. Advances in GPS receiver technology have made it possible to have a 12-channel receiver with the capability of computing the navigation information at a 10 Hz rate, being smaller in size than a credit card, and affordable to an average customer (less than $100).

3.2 Software Receiver

Traditionally, a GPS receiver has the RF and signal processing sections implemented in hardware. The signal processor (usually called the correlator) is generally realized in an Application Specific Integrated Circuit (ASIC). Realizing the correlator in software requires access to the digitized output of the RF section. With the increasing power of microprocessors and in particular digital signal processors (DSP), it has become possible to implement a software-based GPS receiver having only the RF section as the hardware part [Ledvina et al., 2003]. Since the data is processed in software, a modification to the existing processing algorithms involves changing and recompiling of the source code as opposed to the change of the hardware design of the ASIC [Akos et al., 2001].

A software receiver can be customized more easily than a hardware receiver and is a useful research tool to analyze the effect of the following acquisition parameters

1. Predetection Integration Time: The predetection integration time can be varied to determine the amount of acquisition gain obtained and the sensitivity improvement realized [Ledvina et al., 2003].

2. Sampling Frequency: The sampling frequency can be varied to determine the aliasing effect and the processing power requirement. In a real-time software receiver, the sampling frequency also determines the memory requirements.

3. Data Wipe Off: To increase the integration period over the navigation data bit duration, the navigation data bit transition should be determined and the navigation data bit value should be predicted. Different data prediction methods can be implemented and analyzed [Tsui and Lin, 2001].

4. Fine Frequency Estimation: To get a fine estimate of the Doppler, averaging and squaring of the signal are performed using a coarse estimate of the Doppler and code delay. Then a fast Fourier transform (FFT) is performed to obtain a better estimate for the Doppler [Akopian et al., 2002].

3.3   GPS Acquisition

A GPS receiver must detect the presence of GPS signals to track and decode the information for the position computation. A receiver replicates the GPS signal with different PRN codes and performs correlation with the incoming signal. The correlation process yields various peaks that are compared with a detection threshold to test for acquisition success.

The replica signal must match the incoming signal both in code and Doppler. The code phase varies due to the range change between the satellite and the receiver. Doppler variation is due to the relative motion between the satellite and the receiver [Kaplan, 1996]. The role of the acquisition is to provide a coarse estimate of the code phase and the Doppler to the tracking loops. The satellite motion induces a Doppler within ±5 KHz from the GPS L1 frequency [Tsui and Bao, 2000]. User dynamics and clock drift introduce an additional Doppler in the GPS signal. The acquisition Doppler search range should be expanded to include these uncertainties to enable proper acquisition. The code phase search range extends from 1 to 1023 chips (of the C/A-code). The acquisition process searches the signal for a particular value of the code phase and Doppler frequency over a certain period of time called the predetection integration time. The acquisition time is determined by the predetection integration period and the number of cells (obtained from code phase and Doppler range) to search. The GPS receiver can compute visible satellites from approximate knowledge of the receiver position, the GPS time and the almanac which reduces the number of satellites to be searched and speeds up the TTFF.

There have been various acquisition methods developed to acquire GPS signals and a few of them are discussed below.

3.3.1 Time Domain Correlation (cell by cell search)

This is the conventional method for acquisition [Kaplan, 1996]. The search range is divided into cells, wherein each cell represents a particular code delay and Doppler frequency. Correlation is performed in each cell for the predetection integration period and the correlation value is compared against the threshold. If it exceeds the threshold, the satellite is declared as acquired otherwise the search is continued into the next cell. The total number of cells to be searched is given by the number of code delay cells times the number of Doppler bins. This method is simple and best suited for hardware implementation [Tsui and Bao, 2000]. This method performs a sequential search and is time consuming for a software receiver implementation.

3.3.2 Circular Convolution (FFT method)

In this method, the signal is transformed from the time domain to the frequency domain using a Discrete Fourier Transform (DFT) [Van Nee and Conen, 1991]. This method uses the correlation property of the Fourier transform. The property states that the correlation of two sequences in the time domain is the same as the inverse Fourier transform of the convolution of the Fourier transform of the two sequences. For a particular Doppler bin, the correlation of the two sequences performed at all code phase shifts is the same as the inverse Fourier transform of the product of the Fourier transform of the two sequences. Thus, this method reduces the acquisition search range to one-dimension.

The cells are searched in parallel by taking the FFT of the incoming and the local signal which reduces the acquisition time. The steps involved in this scheme are [Van Nee and Conen, 1991]:

1. Collect the sampled IF signal for the desired coherent integration period: $x(t)$

2. Take the FFT of the input signal: $X(F)$

3. Generate the local PRN code for the same coherent integration period and modulate it with the carrier (IF + desired Doppler) and sample it at the same sampling frequency: $y(t)$

4. Take the FFT of the local signal: $Y(F)$

5. Perform convolution in the frequency domain: $Z(F) = (\text{conjugate } X(F)) * Y(F)$

6. Transform the convoluted signal in the time domain: $z(t) = IFFT(Z(F))$

7. Compute the absolute value of the signal $z(t)$, where $z(t)$ represents the correlation of the input signal with the local signal for that Doppler and all possible code phase shifts.

8. Find the peak of the absolute value of $z(t)$ and compare it against the noise threshold. If the peak is greater than the detection threshold, a signal is present. The detection threshold gives an indication of the noise power present. The computation of the detection threshold is explained in Section 5.2. If a signal is

not detected, the procedure is repeated for all possible Doppler values. The detection threshold is optimally based on the noise spectral power density and the allowable probability of false acquisition.

3.3.3 Modified Circular Convolution

This method is same as the circular convolution method (discussed in the Section 3.3.2) except for the length of the FFT which is reduced by half [Tsui and Bao, 2000]. The C/A-code and P-code are transmitted in phase quadrature with each other on the L1 frequency. Hence most of the C/A-code information is contained in the in-phase part of the GPS spectrum. The second half of the spectrum contains little signal information. Hence, this method takes only half the spectrum and performs the correlation [Tsui and Bao, 2000]. The use of half of the spectrum results in a lower number of FFT points. This reduces the FFT processing time and the acquisition time. There is a loss of 1.1 dB determined from simulation analysis, which is due to a loss of the signal information in the other half of the GPS spectrum [Tsui and Bao, 2000].

3.3.4 Delay and Multiply Approach

In this method, the frequency information is eliminated in the input signal and only a code delay has to be searched [Spilker and Parkinson, 1996]. The input signal is multiplied by the delayed version of itself, which eliminates the frequency information but at the same time converts the PRN sequence into a new code. Thus autocorrelation and cross-correlation of the new code have to be performed to determine the code delay.

The problem with this method is that the noise is raised when the input signal is multiplied with its delayed version [Tsui and Bao, 2000]. This method is not useful for acquiring weak signals and hence is not suitable for high-sensitivity receivers.

3.4 Acquisition Detector

The correlation process in acquisition yields correlation peaks. The correlation peak should be above the noise level in the acquisition process to allow the signal to be detected. Noise power computation is an important step in the acquisition process. It is then used to compute the detection threshold. The detection threshold is the minimum value which the correlation peak should exceed for the acquisition process to declare the signal as acquired [Ward, 1996]. An acquisition detector is used to determine the presence of the signal. Most GPS receivers use a multiple trial (M of N / Tong detector) approach compared to a single trial (Binary detector) approach [Kaplan, 1996].

In the binary detector the specified false detection probability along with the noise spectral power are used to determine the threshold. If the correlation value is larger than the threshold, the signal is declared as present [Ward, 1996].

The M of N detector takes N envelopes and compares them to the threshold of each cell. If M or more exceed the threshold, then the signal is declared as present. If not, the signal is declared as absent and the search is repeated for the next cell [Kaplan, 1996].

The Tong detector makes use of an up/down counter to keep a count of the number of times the correlation value has exceeded the threshold. A minimum value of the counter needs to be determined above which the Tong detector declares the signal as present. This value is usually determined by simulations and is a trade off between the search speed and the false detection probability [Spilker and Parkinson, 1996]. There is a limit on the number of times a particular cell is searched before declaring the signal as absent. For weak signals the minimum value of the counter should be kept higher compared to that for strong signals.

3.5   Fine Frequency Estimation

The acquisition process gives a coarse estimate of the Doppler frequency. The tracking loop bandwidth is usually a few Hertz and hence the Doppler frequency estimate should be within the bandwidth of the tracking loop. To obtain a fine estimate of the Doppler frequency, the coherent integration period has to be increased.

The Doppler frequency resolution is given by the inverse of the coherent integration time [Tsui and Lin, 2001]. To obtain a 1 Hz resolution, the coherent integration time has to be one second. The coherent integration time is limited by the navigation data bit transition instant and the Doppler frequency variation. The navigation data bit transition imposes a limit of 20 ms for the coherent integration. The Doppler frequency variation over the coherent integration period causes the correlation value to be multiplied by the sinc signal [Ward, 1996]. This reduces the total correlation value and may even cancel the correlation value. Hence a coarse estimate of the Doppler frequency is determined first to allow fine frequency estimation.

Tsui and Lin [2001] proposed a method to obtain the fine estimate of the Doppler. The phase of the residual signal after carrier wipe-off (using a coarse Doppler estimate) is determined at two instants. The phase difference between the two instants over the time period between the instants gives the fine estimate of the Doppler.

Akopian et al. [2002] developed a method to obtain a fine resolution of the Doppler frequency. A coarse estimate of the Doppler is determined using standard acquisition techniques (discussed in Section 3.3) and is used to wipe off the carrier. The resulting samples are squared to remove the navigation data and are integrated over the desired integration time. A FFT is performed to get a fine estimate of the Doppler. The integration time decides the resolution of the Doppler.

3.6   Weak Signal Acquisition

The navigation data bit duration puts a limit on the coherent integration period. This limit puts a constraint on the processing signal gain in the acquisition process which determines the GPS signal level that can be acquired [Ward, 1996]. To acquire weak signals the predetection integration time has to be extended beyond 20 ms. A method of achieving this is to perform coherent integration for 20 ms and non-coherent integration for the desired duration [Choi et al., 2002]. Non-coherent integration squares and sums the signal across the coherent integration periods. This allows for a coherent integration time to be less than 20 ms and a predetection integration time beyond 20 ms. Non-coherent integration introduces a squaring loss which can be reduced by multiplying the adjacent coherent integration samples over the desired period [Chansarkar, 2000].

Lin et al. [2002] proposed an incoherent integration scheme to reduce the squaring loss present in non-coherent integration. In this scheme, the absolute amplitudes of the coherent integrations are summed up instead of squaring before summation which reduces squaring loss. Multiple thresholds for detection with different coefficients based on the false detection probability were chosen to compensate for the power loss during the correlation due to the Doppler frequency mismatch and the code phase transition.

For a high-sensitivity GPS receiver, the desired acquisition sensitivity is -180 dBW. The nominal noise spectral density in a GPS receiver is usually -204 dBW-Hz. For a precorrelation bandwidth of 2 MHz, the noise contained in this bandwidth is -204 + 63 = -141 dBW and hence the SNR at this sensitivity is -180-(-141) = -39 dB. To detect the signal, the required SNR is about 14 dB and hence the correlation process should provide a gain of 14-(-39) = 53 dB. In addition to this, there are receiver implementation losses such as quantization loss, frequency mismatch loss, code delay mismatch loss and RF bandwidth loss, which together can be around 2 dB [Ray, 2003]. Hence the required correlation gain is 53+2 = 55 dB. The coherent integration gain is given by $10\log_{10}$ (1023 * coherent integration period). For example, with 20 ms coherent integration, the gain obtained is 43 dB. So the SNR before non-coherent integration is -39-2+ 43 = 2 dB. Non-coherent integration introduces a squaring loss of 3 dB for an SNR of 2 dB [Ray, 2003] and hence the required gain is 14-2-(-3) = 15 dB. Non-coherent integration should provide for this 15 dB gain. The non-coherent integration period is given by $10^{(gain\ needed/10)}$, i.e. $10^{(15/10)}$ = 32 times the coherent integration of 20 ms. In this manner, the coherent and non-coherent integration periods can be chosen. Thus each cell will take 640 ms (32 * 20 ms) to be searched. To reduce the TTFF, a parallel correlator can be used and multiple cells can be searched at the same time.

3.7 Satellite Search

The TTFF can be reduced by properly organizing the order in which the satellites are searched. If almanac (less than 4 month old), time (accurate up to 10 min) and user position (accurate up to 100 km) are available, then the list of visible satellites can be computed and satellites can be searched in descending order of their elevation angle. If any of this information is not available, the list of visible satellites cannot be computed. In that case, a sky search has to be performed. The order in which the satellites are searched during a sky search will determine the TTFF [Pietilä and Syrjärinne, 2000]. Hence, the order has to be carefully chosen to reduce the TTFF. A few methods to select the order of satellites to be searched are:

1. Sequential order from PRN number 1 to 32 [Ward, 1996].

2. Random selection [Kaplan, 1996].

3. Satellites in orbital plane-wise [Pietilä and Syrjärinne, 2000].

4. Statistical distance measure between satellites and building up a tree with different levels such as 4, 8 12 and so on [Pietilä and Syrjärinne, 2000].

3.8 GPS Tracking

The tracking process follows from the acquisition process and keeps the lock on the signal and generates the measurements. The receiver should keep track of the carrier and code phase of the incoming satellite signals. The tracking loops consist of a loop filter, discriminator and either a voltage controlled oscillator (VCO) or a numeric controlled oscillator (NCO). A VCO or NCO generates the local signal to match the incoming signal. The difference between the incoming signal and a local signal is averaged in the loop filter and then passed to the discriminator to determine the error. The discriminator output gives an error which is fed back to the VCO/NCO to correct the generation of the local signal [Kaplan, 1996]. The tracking process also detects for the loss of lock on the satellite signal. If there is a loss of lock, the signal has to be reacquired through the acquisition process.

An FLL outperforms a PLL under dynamic stress and RFI conditions while a PLL gives better measurement accuracy than the FLL [Kaplan, 1996]. An FLL-assisted-PLL solves the GPS receiver designer's dilemma when faced with the need for both the dynamics robustness of the FLL plus the accuracy performance of the PLL [Ward, 1998]. A well-designed code tracking DLL will track at considerably lower levels of C/No than the carrier tracking loop in an unaided (stand-alone) GPS receiver. Since both the code and carrier tracking loops must successfully track their respective signals in order for the unaided GPS receiver to operate, it is sufficient to analyze only the weaker (carrier) tracking loop to determine the overall receiver tracking threshold (effective C/No below which the carrier tracking is no longer successful).

3.9   Acquisition Performance Parameters

Acquisition is performed for each satellite over the code phase and Doppler range. A predetection integration time is a combination of the coherent and non-coherent integration time. The coherent integration time algebraically adds up the signal and noise while the non-coherent integration time performs the absolute sum of the signal and noise across coherent integration periods [Tsui and Lin, 2001]. The predetection bandwidth is obtained as the inverse of the coherent integration period. It determines the number of Doppler bins to be searched over the complete Doppler search range and decreases as the coherent integration period increases [Ward, 1996]. An increase in the coherent integration period accumulates more signals and allows the noise to average out thereby increasing the SNR. However, an increase in the coherent integration period reduces the predetection integration bandwidth thereby increasing the number of Doppler bins [Tsui and Lin, 2001]. This increases the number of cells to be searched which along with an increase in the coherent integration time increases the total acquisition time at the cost of a higher SNR. The optimum value of the coherent integration time should be chosen to meet the gain and time requirements of the acquisition.

Non-coherent integration is used to increase the gain while keeping the coherent integration time as small as possible [Choi et al., 2002]. It adds up the signal and noise across the coherent integration periods, thereby increasing both the signal and noise power. The amount of increase in the signal power is linear while the noise power does not add up linearly. Noise is white Gaussian in nature and varies across coherent integration periods [Kaplan, 1996].

The quality of the acquisition is usually judged by its mean acquisition time, acquisition sensitivity (cold, warm and hot starts), and its false alarm and missed detection probabilities [Lin et al., 2002]. The acquisition performance is characterized by the acquisition gain, the mean acquisition time and memory requirements.

Acquisition gain is the ratio of the correct signal peak to the detection threshold and depends on the length of the predetection integration period. The gain is lower in the weak signal environments, under multipath and in interference conditions [Tsui and Bao, 2000]. It gives an indication of the allowable increase in the noise power before the failure of the acquisition process. It is usually expressed in terms of decibels (dB) or SNR.

The mean time for acquisition is determined by the predetection integration period, the number of Doppler bins and the code phase cells. The total Doppler search range is determined by the satellite motion, the receiver clock offset and the receiver dynamics [Kaplan, 1996]. The acquisition search range can be reduced if the approximate values of the Doppler and code phase are known a priori. Acquisition has to be as quick as possible to allow the GPS receiver to provide a navigation solution almost instantly after power on.

Embedded systems have limited the amount of memory available for processing so the available memory has to be efficiently distributed among various GPS processing blocks. In this context, the memory consumed by the acquisition process becomes significant. The acquisition process should utilize as few memory locations as possible. This can be achieved by using a smaller coherent integration time or averaging of samples considered for detection.

CHAPTER 4: RFI: EFFECTS AND MITIGATION STRATEGIES

This chapter discusses various interference types and their effects on the GPS signal processing. Different jamming techniques and interference mitigation strategies are also discussed.

4.1   Interference Signals

RFI is a major source for degradation of the GPS accuracy and reliability. Since there are other sources of errors which further degrade GPS accuracy, this makes RFI mitigation more difficult. GPS satellites and users are mobile which make it difficult to integrate the signals over long periods of time to average out the effects of noise. Satellite and user motion introduce Doppler effects, slow power fluctuations (due to changes in the effective antenna gain and path loss) and fast power changes (due to multipath fading, blockage and shadowing) [Heppe and Ward, 2003]. A Doppler fluctuation makes it difficult to distinguish between user motion and receiver clock drift. Power fluctuations make it difficult to determine the thresholds for acquisition and tracking. Atmospheric errors introduce range and range-rate errors.

RFI and jamming are two major concerns in using GPS to provide a reliable solution [Kaplan, 1996]. Unintentional interference can be caused by RF transmitters, harmonics of ground transmitters, radar signals and accidental transmission of signals in the wrong frequency band [Spilker and Parkinson, 1996]. The signals, or the harmonics of the signals, near the GPS frequencies (L1 and L2), are potential sources of interference.

47

Interference can also be caused by ionospheric scintillation and evil waveforms transmitted by the GPS satellites themselves [Geyer and Fraizer, 1999]. Pulsed interference can result from radar signals in nearby frequency bands which are not properly filtered [Littlepage, 1999]. Table 4.1 summarizes various types of RFI with a few interference sources shown in Figure 4.1.

Table 4.1: Types of RFI and possible sources [Kaplan, 1996]

Type	Typical source
Wideband-Gaussian	Intentional noise jammers
Wideband phase/frequency modulation	Television transmitter's harmonics or near-band microwave link transmitters
Wideband-spread spectrum	Intentional spread spectrum jammers or near-field of pseudolites
Wideband pulse	Radar transmitters
Narrowband phase/frequency modulation	AM stations transmitter's harmonics
Narrowband swept continuous wave	Intentional CW jammers or FM stations transmitter's harmonics
Narrowband continuous wave	Intentional CW jammers or near-band unmodulated transmitter's carriers

CW interference can be either a pure tone or a narrow band modulated signal such as AM or FM [Macabiau et al., 2001]. It distorts the signal spectrum and affects the carrier tracking loop. A carrier tracking loop will lock onto the interference frequency for a pure tone signal (provided the CW power level is considerably high) generating erroneous carrier phase and Doppler measurements. Broadband noise increases the amount of noise in the GPS spectrum without distorting the signal spectrum [Heppe and Ward, 2003]. Swept CW interference is more damaging than CW interference because it can cover multiple Doppler frequencies and affect more than one receiver channel at the same time.

Pulse interference can cause malfunctioning of the AGC which affects the tracking loops [Hegarty et al., 2000].

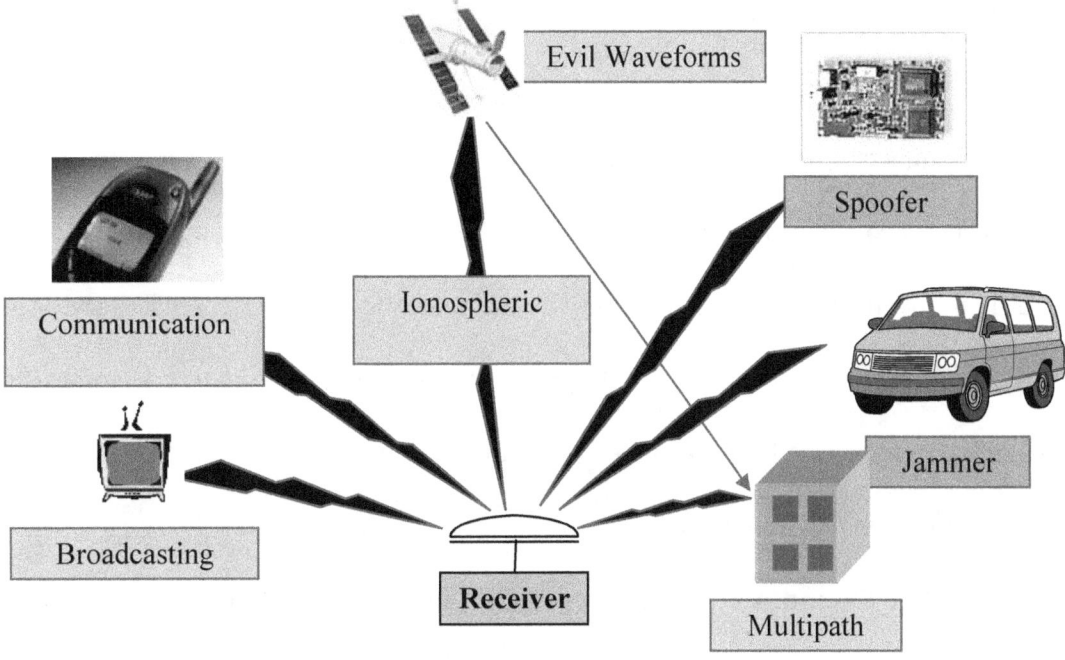

Figure 4.1: Some sources of RF interference

An AM and FM radio broadcast transmitter's high-order harmonic emissions fall close to the GPS L1 frequency and cause interference. RFI likelihood is considered minimal for AM radio broadcasts since the harmonic order (985) is high [Erlandson and Fraizer, 2002]. For an FM broadcast the harmonic order (15 to 18) is lower and the maximum effective isotropic radiated power (EIRP) is higher (50 to 60 dBW) [Macabiau et al., 2001]. Analog TV broadcast maximum EIRP limits are higher than FM and the harmonic orders are lower (2 to 9 for RFI signals within 2 MHz of the GPS L1 frequency) and the predicted minimum separation radius exceeds 100 miles [Buck and Sellick, 1997]. Thus FM and TV signals will cause interference in the GPS receiver. Buck and Sellick [1997] analyzed the effects of the harmonics of the TV signals interfering with GPS frequencies and they were found to be in the L1 signal spectrum causing a non-linear effect. The

49

strongest suspected interference signal was at 525.25 MHz (video carrier of a local TV station). Thus the 1575.75 MHz signal was the third harmonic of the local station video carrier. The GPS L1 frequency divided by three is 525.14 MHz and the transmitted TV signal's lower side band suppression was at 524.50 MHz thereby allowing full power at this frequency. This jump in power would produce a high level of interference resulting in a reduced SNR. Filters were found to be effective in eliminating these interference signals by having high attenuation for the undesired signals. The TV and Air Traffic Control (ATC) frequencies have high transmitter powers and their harmonics fall in the GPS L1 frequency band. Table 4.2 lists the various TV and ATC frequency harmonics falling in the GPS frequency band.

Table 4.2: TV and ATC harmonics in GPS frequency band [Buck and Sellick, 1997]

Harmonic	Band (MHz)	Usage
2	787.199 – 787.222	Broadcasting
3	524.799 – 525.481	Broadcasting
4	393.599 – 394.111	Fixed/mobile
5	314.879 – 315.289	Fixed/mobile
6	262.400 – 262.741	Fixed/mobile
7	224.914 – 225.206	Broadcasting
8	196.800 – 197.055	Broadcasting
9	174.933 – 175.160	Broadcasting
10	157.440 – 157.644	Broadcasting
11	143.127 – 143.313	Fixed/mobile
12	131.200 – 131.370	ATC
13	121.107 – 121.265	ATC
14	112.457 – 112.603	VOR
15	104.960 – 105.096	Broadcasting

The best protection for a GPS receiver is to use RF filtering to exclude the unwanted interference. Spurious transmissions from RF transmitters in the GPS frequency band should be measured to allow its suppression [Johannessen et al., 1990].

4.2   Interference Effects

RFI has the same effect on GPS acquisition or tracking as signal blockage, foliage attenuation, ionospheric scintillation and multipath, which is to reduce the C/No for all the GPS signals. A jammer reduces the SNR of the GPS signals affecting acquisition and tracking of the signals in the GPS receiver. Spoofing is another form of interference which transmits a stronger version of the GPS signal to capture the receiver loops and fool the receiver [Heppe and Ward, 2003]. Pseudolites operating at close range to a receiver can jam the GPS receiver. The primary aspect of the GPS architecture that makes it vulnerable is the low power of the signal that is actually below the noise floor until it is de-spread with an appropriate PRN code. The RFI effect depends on the details of receiver design, especially the front-end bandwidth and early-late spacing in the discriminator [Macabiau et al., 2001]. It has a different effect on the code tracking accuracy than it does on some other aspects of the GPS receiver [Geyer and Fraizer, 1999]. Several types of perturbations like the thermal noise, atmospheric disturbances, multipath and interference can affect the GPS signal. Geyer and Fraizer [1999] conducted tests on a C/A-code receiver for the FAA to determine the vulnerability of the GPS receivers to RFI. This allowed the FAA to establish interference standards for GPS receivers used in civil aviation. These tests were focused on the C/A-code receiver's tracking degradation and loss of lock under different interference conditions. The GPS was found vulnerable to very high frequency (VHF) transmissions and CW interference.

RFI detection should be given high priority because it provides an instantaneous warning of the potential loss of GPS integrity. It can be detected using a jamming-to-noise (J/N) power ratio meter [Kaplan, 1996]. The J/N meter is implemented in the AGC of the GPS receiver front-end. This meter keeps a check on the thermal noise level and any signal different from it, is detected as the presence of the interference signal.

The C/No for a SV signal without interference is termed as unjammed C/No [Kaplan, 1996]. The difference between the unjammed C/No and the acquisition or the tracking threshold gives an indication of the possible interference tolerance and is termed as effective C/No. The unjammed C/No and the effective C/No are used to compute the maximum jammer-to-signal (J/S) level at the receiver input from which the RFI power can be determined. The unjammed C/No depends upon the GPS receiver parameters and is computed from Equation (4.1) [Kaplan, 1996].

$$C/No = Sr + Ga - 10\log(kTo) - Nf - L \ (dB - Hz) \qquad\qquad 4.1$$

where

Sr	is the received GPS signal power (dBW),
Ga	is the antenna gain towards the SV (dBic),
$10\log(kTo)$	is the thermal noise density (dB-Hz) $\cong$ -204 dBW-Hz,
k	is the Boltzmann's constant (watt-sec/K) = 1.30×10^{-23},
To	is the thermal noise reference temperature (K) = 290 K,

Nf   is the noise figure of the receiver (dB), and

L   is the implementation loss plus ADC loss (dB).

Signal information is lost during conversion of the signal from analog to digital by the ADC which is referred to as the ADC loss. The level to which the unjammed C/No is reduced by the RFI is called the equivalent C/No power density ratio. The equivalent C/No power density ratio is related to unjammed C/No and J/S as given by Equation (4.2) [Kaplan, 1996].

$$[C/No]_{eq} = ((C/No)^{-1} + (J/S)/QR))^{-1} \quad \text{(power ratio)} \qquad 4.2$$

where

C/No   is the unjammed carrier-to-noise power in a 1 Hz bandwidth expressed as a ratio,

J/S   is the jammer-to-signal power expressed as a ratio,

R   is the GPS PRN code chipping rate (chips/sec), which

is 1.023×10^{6} chips for the C/A code and 10.23×10^{6} chips for the P code, and

Q   is the spread spectrum processing gain adjustment factor, and

is 1 for narrow band jammer, 1.5 for wide spread spectrum jammer and 2 for wideband Gaussian noise jammer.

Equation (4.2) can be expressed in terms of dB-Hz (Equation (4.3)) and rearranged to obtain J/S (Equation (4.4)) [Kaplan, 1996].

$$[C/No]_{eq} = -10\log[10^{-(C/No)/10} + 10^{(J/S)/10}/QR](dB-Hz) \qquad 4.3$$

$$J/S = 10\log[QR(10^{-([C/No]eq)/10} - 10^{-(C/No)/10})](dB) \qquad 4.4$$

For an C/A-code receiver, Sr = -159.6 dBW and assuming the antenna has unity gain toward the SV (Ga = 0), a noise figure of 4 dB and an implementation loss of 2 dB, then the unjammed C/No is 38.4 dB-Hz. For Q=2, and assuming an equivalent C/No threshold of 28 dB-Hz, the J/S = 34.7 dB [Kaplan, 1996]. This tolerance looks good in terms of dB but when converted to the actual signal power it is just 3 pW. The RF transmitter transmits signals with high power levels (in terms of Watts) and hence the harmonics of these signals will have power levels greater than 3 pW. This will result in jamming of the GPS receiver and hence RFI detection and mitigation is important in a GPS receiver.

A number of techniques have been designed to increase the robustness of GPS receiver to RFI signals [Littlepage, 1999]. RFI can be mitigated at various stages of the GPS receiver from the instant of receiving the GPS signals by the antenna to the position computation instant. RFI signals will have full effect when the interference signal is unobstructed and the antenna provides adequate gain to the signal.

4.3   GPS Jammer

Jamming or spoofing is a form of an intentional interference. Jamming can be of the form of signal denial (which prevents acquisition and tracking of the GPS signal) or signal deception (which fools the receiver to mistake an interference signal as the GPS signal) [Heppe and Ward, 2003]. A simple jammer can be constructed to generate a CW, AM or FM interference. Jammer design becomes complex when a range of jamming types are to be generated from the same source. A brute force jamming method introduces broadband noise or CW jamming to prevent lock on the satellites. An intelligent jammer uses different types of interference signals to attack the GPS receiver. Pulse signals can be used to attack the AGC and tracking loops and a swept CW signal can capture all receiver channels by exploiting C/A-code spectral lines. The frequency of these signals can be varied to reduce the chance of jammer detection [Cutright et al., 2003]. The effect of jamming on the RF front-end is to create additional spurious signals and tones which disrupt receiver operations. Jamming forces the low noise amplifier (LNA) into saturation and yields spurious signals. It forces the AGC to respond at incorrect operating points by exploiting the tracking loop constant and signal suppression in the ADC [Heppe and Ward, 2003].

Interference signals within the GPS C/A-code bandwidth are termed as Narrowband interference [Kaplan, 1996]. Narrowband interference signals are potentially more dangerous than expected due to the line spectrum of the Gold codes used for ranging. They could coincide with a strong spectral line which results in a larger residual line in the carrier tracking loop [Macabiau et al., 2001]. The CW interference effect on tracking can be detected due to the Doppler variation between the satellite and a user. A broadband RFI averages the spectral lines and causes an asymptotic effect like a sinc function [Kaplan, 1996]. The effective interference signal power at the receiver can be determined using a link budget knowing the power transmitted by the interference source and is calculated using Equation (4.5) [Heppe and Ward, 2003].

$$\text{Received RFI} = \text{EIRP - Path loss + Antenna Gain} \qquad\qquad 4.5$$

The EIRP represents the effective power transmitted by the source. The signal is weakened by the path loss as it propagates through space. The path loss can be calculated by Equation (4.6).

Path Loss (dB) $= 31.8 + 20 \log (f) + 20 \log (\rho)$ 4.6

where

    f   is the frequency in MHz, and

    ρ is the range in km.

The antenna gain impacts the amount of interference received by the receiver. It can be designed to provide attenuation in the direction of the interference signal provided the direction is known [Bond and Brading, 2000].

GPS jammers are available commercially at low cost compared to the GPS receivers and these jammers can transmit different types of waveforms. The most difficult interference waveform for the GPS receivers to mitigate is broadband noise [Kaplan, 1996]. Since the GPS signal is a spread spectrum signal; a narrow band pass filter in the receiver front-end that removes the jamming signal will remove only a small portion of the GPS signal while removing the narrow band jamming signals. The GPS signal has a low power level which makes it vulnerable to jamming over long distances. Brown et al. (1999) analyzed the effect of a single 4-Watt CW jammer on the GPS acquisition in a GPS receiver. The results illustrated in Figure 4.2 deny the C/A-code acquisition for a J/S ratio of 22 dB. GPS tracking was jammed when the jammer was close to the receiver (1 km) while acquisition was jammed over a larger distance. Airborne users are more affected by a ground-based jammer than ground-based GPS users as they have a direct line-of-sight (LOS) to the jammer over significant distances (up to 145 km) as shown in Figure 4.2.

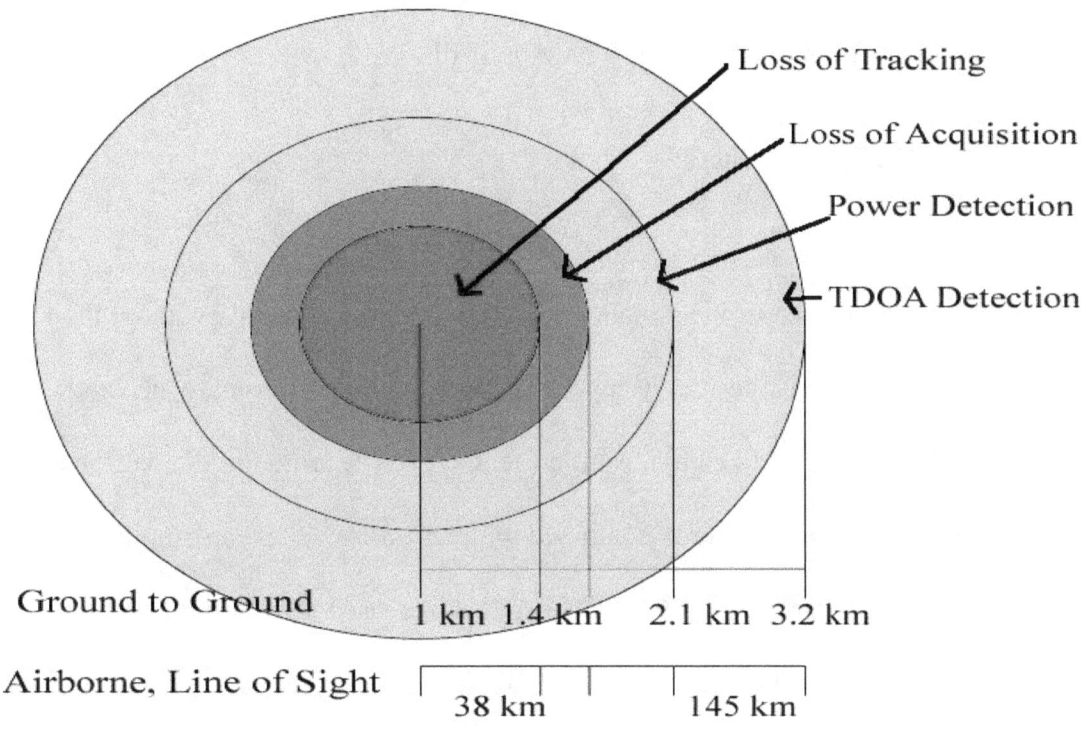

Figure 4.2: Effective range of single 4-watt GPS jammer [Brown et al., 1999]

A GPS receiver can be successfully jammed by generating interference signals that match the Doppler and by offsetting the jammer to a maximum spectral line of the C/A-code. The correlation process of a CW line and a PRN code will spread the CW line, but the mixing process at certain strong C/A-code lines results in the RFI line being less suppressed than other frequencies [Johnston, 1999]. The net result is that CW interference can leak through the correlation process at this strong line. Johnston [1999] performed experiments on CW and swept frequency interference using a commercial off-the-shelf (COTS) NavSymm/Navstar XR5-M 12 channel GPS receiver used in conjunction with Navstar software version 3.7. DGPS was used to remove the effects of

Selective Availability (SA) to obtain the error due to the jamming. The results showed that CW jamming gave a maximum position error of 23 km while the swept frequency gave an error of 220 m for a J/S ratio of 35 dB.

4.3.1 Simple Jammers

A simple jammer is one which generates interference signals without the knowledge of the GPS receiver design [Heppe and Ward, 2003]. It usually generates one type of interference signal like broadband noise, CW, etc. A brute force broadband noise jammer uses a simple noise source combined with an up-converter, an amplifier and an antenna [Spilker and Parkinson, 1996]. This causes increased tracking jitter, cycle clips and loss of lock. It is effective over short ranges and the goal is to deny the GPS signal. Acquisition can be disrupted at 3-4 times the range at which tracking is lost [Heppe and Ward, 2003]. It is low cost and affects all the receivers within its effective range. The jammer power should be high to affect over a long range but it also increases the probability of detection.

A CW jammer is simply a tone frequency generator at GPS frequencies. Its purpose is to capture the carrier tracking loop of the receiver and to mislead the user. A high power CW signal can cause the tracking loop to appear as stable but will generate erroneous measurements [Johannessen et al., 1990]. It is effective over medium ranges, is low cost, and affects all the receivers within its range. The disadvantage is that it is easier to detect than a broadband noise jammer and can be filtered out prior to correlation during acquisition or tracking [Burns et al., 2002].

4.3.2   Intelligent Jammers

A jammer designed with the knowledge of the GPS receiver architecture is called an intelligent jammer [Heppe and Ward, 2003]. An intelligent jammer generates complex waveforms like pulse signals or sweep CW signals making the RFI detection difficult. For example a pulsed noise jammer is more complex in design than a CW jammer. The key parameters in design are the pulse duration, duty cycle and amplifier efficiency. The intention of a pulse jammer is to disrupt tracking and data demodulation. The pulses are designed to match the time constant of the AGC and this can give 10's of dB advantage to the jammer [Hegarty et al., 2000]. A pulse jammer requires less power since the signal is not continuous which makes it more difficult to detect. Its disadvantage is that it requires a high power amplifier and the knowledge of the receiver design to affect the AGC [Heppe and Ward, 2003]. Knowledge of the receiver architecture can help design a pulse jammer to keep the receiver constantly in acquisition mode even though the jammer is never constantly on.

A swept CW (chirped jammer) allows the capture of the carrier tracking loops for all signals despite the Doppler difference [Johnston, 1999]. This jammer can be used to attack rapidly moving vehicles and can be turned off to save power after the receiver has been jammed [Heppe and Ward, 2003]. Knowledge of the receiver's reacquisition time is used to determine the on/off period of the jammer. An interference signal close to the receiver's IF is difficult to isolate using hardware techniques. A frequency hopping jammer is similar to a CW or pulse jammer whose frequency is varied in steps over the desired bandwidth of the signal. The frequency variation makes it difficult to detect the interference signal [Heppe and Ward, 2003]. The frequency variation rate can be changed to make it more difficult to detect.

4.3.3 Spoofers

Spoofing is an intentional jamming signal similar in nature to the GPS signal. It can be of two types: navigation and data link signals [Kaplan, 1996]. Navigation spoofers transmit false GPS signals at a high power to fool the receiver and capture it. This will introduce large errors in the GPS measurements and will provide an erroneous navigation solution. These false GPS signals should be carefully designed to be able to fool the receiver [Heppe and Ward, 2003]. Data link spoofers capture the navigation data signal and require about an extra 10 dB of power compared to the navigation spoofers. The navigation data is spoofed to provide erroneous information about the range accuracy, GPS time, satellite positions and their health. Both the spoofing signals can be combined to severely degrade GPS solution [Heppe and Ward, 2003].

Spoofing requires capturing of the tracking loops and eliminating the actual GPS signals. Therefore, spoofer must track the receiver trajectory and generate signals to match the GPS signals. Once the receiver starts tracking the spoofed signals, their power can be increased to eliminate the GPS signals and then the erroneous information can be transmitted to fool the receiver. The spoofer should track the GPS signal and the receiver's motion to have a tight closed loop signal generation [Heppe and Ward, 2003]. Receiver autonomous integrity monitoring (RAIM) and false detection and identification

(FDI) methods in a GPS receiver can be used to isolate the erroneous GPS measurements when there are more than four measurements. Hence the spoofer should spoof all the satellite signals to throw the receiver off its trajectory [Spilker and Parkinson, 1996].

4.3.4 Pseudolites

The GPS signal power varies about 20% between the satellites at the horizon and at high elevations [Kaplan, 1996]. The antenna gain pattern is usually designed to ensure the GPS signal strength does not vary much between the satellites. The use of pseudolites to augment GPS satellites gives rise to the near-far problem [Madhani and Axelrad, 2001]. A pseudolite signal has a large power variation when it is close to the GPS receiver. A pseudolite designed to provide a GPS signal level at a distance of 50 km would provide about 60 dB more power at a distance of 50 m from the receiver [Madhani and Axelrad, 2001]. This near-far problem can be overcome using different methods classified into three categories namely signal pulsing, frequency offsets and the use of different PRN codes. Signal pulsing consists of transmitting the pseudolite signal in terms of the pulse which decreases the average power of the pseudolite signal. The pseudolite signals can be transmitted at a frequency outside the GPS frequency band which requires changes to be made in the receiver front-end [Ndili, 1994].

Madhani and Axelrad [2001] developed a successive interference cancellation approach to overcome the near-far problem in the pseudolites. This approach identifies and eliminates the strongest component in the received signal. The next strongest component is then identified and cancelled. This is repeated until all the signals are separated. The strongest signal is removed first since it is easy to acquire and its removal gives the most benefit for the remaining weak satellite signals.

4.4 RFI Mitigation Methods

GPS has an advantage over narrow-band navigation systems with respect to unintentional interference due to the following reasons. The GPS signals are spread-spectrum signals and receiver design techniques can reduce the effect of most of the interference signals [Kaplan, 1996]. A GPS navigation solution is usually over determined and RAIM/FDI methods can be used to isolate erroneous information [Spilker and Parkinson, 1996]. The correlation process in a GPS receiver de-spreads the GPS signal and spreads any interference signal present which reduces the interference power and provides some protection against interference signals. However, a high power interference signal can distort the correlation peak or give rise to the correlation peak at incorrect estimates [Heppe and Ward, 2003]. The main strategy for any interference mitigation method is to eliminate the interference signal or reduce the interference signal to white Gaussian noise, so that there is only an increase in noise without distortion of the GPS signal spectrum [Cooper and Daly, 1997].

4.4.1 RF Filtering

GPS receivers operating close to RF broadcasting devices like TV stations, high power transmitters will suffer from out-of-band interference [Escobar and Harper, 2001]. To eliminate out-of-band interference high performance RF filters are used between the antenna and the receivers. These RF filters are required to have a sharp cut-off outside the GPS bandwidth, low loss in the pass band and high rejection in the stop band. Superconducting Technology Inc. constructed a set of HTS filters for the L1 and L2 frequencies along with a cryogenically cooled LNA. The use of cryogenic technology decreases the loss in the filter improving its performance. This improvement in the performance allows for the use of normal filters after the LNA without much affect on the Noise Figure (NF) of the receiver front-end [Escobar and Harper, 2001]. The development of surface acoustic wave (SAW) filters has reduced the size, weight, cost and cooling requirements for the filters.

4.4.2 Adaptive Antenna Array

A GPS antenna captures the GPS signal and feeds it to a receiver. Any interference signal in the GPS frequency band is also picked up by the antenna [Bond and Brading, 2000]. RFI can be eliminated by providing a zero antenna gain in the direction of the interference signal. Antenna gain pattern cannot be modified for a single element antenna without changing its orientation [Kunsyz, 2001]. It can be varied by using an array of antenna elements. An adaptive antenna array works on this basic concept of providing a zero gain to the interference signal. The adaptive antenna arrays can be either fixed rejection pattern antenna (FRPA) or CRPA. Both these antenna arrays consist of an adaptive processor to combine signals from different antenna elements. The manner in which the signals from the different antenna elements are combined can be varied to change the overall gain pattern of the array.

The FRPA system consists of an array of three conventional antenna elements and a vector processor which contains the adaptive antenna RF circuits and a processor [Littlepage, 1999]. A minimum of three elements are required to determine the interference location in both the azimuth and elevation and to provide a uniform azimuth accuracy throughout the area visible to the antenna. The spacing between the antenna elements is important to avoid directional ambiguities [Bond and Brading, 2000]. The interference direction is determined from the angle of arrival (AOA) and array theory. The DF algorithm checks the results across the antenna pairs to ensure consistency and to trap the errors [Bond and Brading, 2000]. The number of antenna elements in the array determines the number of the interference sources that can be eliminated. FRPAs provide substantial jammer suppression at a relatively low cost but they are vulnerable to the distributed broadnoise jammers [Kunsz, 2001]. This problem can be overcome using spatial filtering (adaptive nulling) provided by the CRPAs [Littlepage, 1999].

Kunysz [2001] designed a compact, dual GPS frequency L1/L2, multi-element CRPA array. Each antenna element consists of an aperture coupled spiral slot array which reduces the mutual coupling between the adjacent elements. This provides better tolerance towards the interference or a multipath signal and helps to reduce the size of the array. This antenna was successfully found to mitigate a wideband jammer and multipath signals in standard surveying, marine and arctic applications.

4.4.3   Interference Localization

RFI can be mitigated provided the direction and nature of the interference signal is known. Interference localization is to determine the direction of the interference signal [Brown et al., 1999]. There are several ways to implement interference localization such as interferometry, TDOA systems, spatial spectrum estimation, phase antenna arrays, etc. The interferometrical approach uses a concept of direction finding to localize the interference source. It employs a group of signal recognition and direction finding equipment to locate the interference source. This technique is useful for locating a small number of jammers but is not practical to locate a large number [Brown et al., 1999].

TDOA techniques can be used to locate a large number of jammers. The time delay of the interference signal reaching the multiple antennas can be used to determine the location of the interference source [Gormov et al., 2000]. Doppler measurements from each antenna are used to determine the satellite motion and to isolate any other signal, but these measurements provide erroneous information when the carrier drifts or is intentionally dithered. The direction finding techniques using the antenna arrays are limited in the number of separate jammers that can be located [Brown et al., 1999].

Many simultaneous measurements are necessary to quickly and accurately locate a large number of jammers and spoofers [Shau-Shinu and Enge, 2001]. A simple method is to use the C/No information from a GPS receiver to determine the jammer location. The satellite signal strength, and the GPS time and location of the receiver can be used to determine the jammer location [Brown et al., 1999]. This method is effective when there is a large variation in the C/No as a function of the distance from the jammer. However, it is less effective in estimating the location of low power interference signals. RFI source location can be estimated using a network of distributed sensors rather than a single sensor [Shau-Shinu and Enge, 2001]. The network approach to locate the RFI source requires no sensor motion and is robust to sensor failures.

Shau-Shinu and Enge [2001] developed an aircraft RFI localization and avoidance system (ARLAS) to reduce the interference in aviation applications. The system uses a GPS antenna mounted on the top of an aircraft to determine the interference location. The SNR of the received GPS signal is calculated by the GPS receiver under different values of roll, pitch and heading which are measured from the gyros. This information along with the vertical gain information of the GPS antenna is used to estimate the direction to the interference source.

4.4.4 AGC as Interference Mitigation Tool

GPS signals received by the antenna have an inherent thermal noise present in them. This thermal noise is determined by the AGC and is used to determine the thresholds for quantization. Bastide et al. [2003] studied the AGC as a tool for interference assessment. An AGC provides an accurate indication of the thermal noise in the receiver. These noise levels can be used to determine the presence of interference. Any interference signal will increase the noise power in the receiver which can be detected using the AGC.

Bastide et al. [2003] devised a Chi-square test to detect the presence of interference using the distribution of the ADC bins.

4.4.5 Pulse Blanking

Pulse interference signals affect GPS receivers depending on its characteristics such as power, duty cycle and pulse width [Hegarty et al., 2000]. They will continue to affect the receiver components even in it's off state because the components have a recovery period to resume their normal operation. These signals tend to saturate the RF stages, the AGC and the ADC in a receiver front-end. Slow AGCs will be severely affected by pulse interference. These AGCs are slow to respond and will incorrectly determine the quantization levels, which will result in improper sampling [Hegarty et al., 2000]. A fast AGC along with an increase in the number of quantization levels will solve this problem. Hegarty et al. [2000] devised a technique to eliminate pulse interference through blanking. In this method, whenever the pulse interference is detected, the ADC outputs a zero thereby eliminating the interference signal. However this introduces a signal loss at the samples where the pulse interference was detected. Perfect blanking for a single strong-pulsed signal will result in SNR degradation of 10log(1-PDCB) where PDCB (pulse duty cycle – blanker) is the duty cycle of the blanking signal. This SNR degradation follows from the fact that when strong pulses are present, blanking completely suppresses the desired signal (20log (1-PDCB) SNR degradation) and thermal noise (10log (1-PDCB) SNR gain) [Hegarty et al., 2000].

4.4.6   Spatial Signal Processing

Navsys Inc. pioneered the first commercial receiver to include spatial signal processing; the high-gain advanced GPS Receiver (HAGR). This receiver uses digital spatial processing to combine signals from the antenna (up to 16) elements. Brown et al. [2000] further enhanced the HAGR to detect the presence of the interference signals and to estimate its direction. The AOA of the interference signal is determined using the cross-correlation between multiple antenna elements. A cross-correlation value higher than the noise power indicates the presence of an additional interference signal. This technique was found capable of detecting CW interference above a -129 dBm power level and a broadband interference above -125 dBm [Brown et al., 2000].

4.4.7   Space Time Adaptive Processing (STAP)

Antenna arrays equipped with a STAP can null multiple RFI signals arriving from different directions without previous knowledge of the interference type or the direction of arrival (DOA) [Moore and Gupta, 2001]. An antenna array with L elements and N taps for each element will capture LN signals and feed it to the STAP. The STAP adjusts the coefficients for each tap and the array element to provide a zero gain in the direction of the interference. For this antenna, there are LN-1 DOF available and hence it can eliminate up to LN-1 independent RFI sources. Moore and Gupta [2001] demonstrated that a wideband RFI source could consume multiple spatial DOF depending on the interference power and bandwidth. The STAP is required to process matrices of order LNxLN and was found to distort the desired GPS signal introducing significant errors in the navigation solution under severe jamming conditions. This problem was overcome using a SFAP [Gupta and Moore, 2001].

4.4.8 Spatial Frequency Adaptive Processing

SFAP is an alternative solution to the STAP whereby signals from the antenna elements are processed in the frequency domain. STAP and SFAP are equivalent if the tap spacing is equal to the sampling interval [Gupta and Moore, 2001]. The SFAP performance can be improved by using a window function to multiply the time domain samples before transforming them into the frequency domain. The window function (e.g. Blackman window) should have low side lobes to provide better performance. The narrowband SFAP does not distort any of the desired signals as observed in the STAP [Gupta and Moore, 2001].

4.4.9 RFI Mitigation in the GPS Correlator

Macabiau et al. [2001] devised a multicorrelator technique for the CW interference detection. A multicorrelator receiver provides correlation values of the incoming signal with several delayed replicas of the same local code in a single tracking channel. CW interference disturbs the in-phase and quadrature correlator outputs of the tracking loop and this effect varies as the spacing between the correlators' changes. This variation in the correlator outputs is used to determine the frequency of the CW interference.

Manz et al. [2000] developed a technique to mitigate the RFI in the PLL. This method can be used only if the user is stationary and has a stable clock. The amount of thermal noise present in the PLL is determined and the variation of the noise power in the PLL is studied. An increase in the PLL noise due to interference is equated with the noise power injected into the PLL. The ratio of the interference power to the C/A-code signal power is determined which indicates the interference power required to jam the PLL. A narrow PLL bandwidth increases the probability of tracking the correlation side lobes. The standard algorithm used to detect the 25 Hz side lobes can be modified to detect the 12.5 and 8.33 Hz side lobes [Tsui and Bao, 2000]. The narrow PLL bandwidth improved the CW interference rejection performance in a GPS receiver.

Cooper and Daly [1997] developed a technique of preprocessing the GPS signals to remove the interference components before passing them to the GPS correlator. The frequency of the interference signal is determined using frequency domain techniques and a PLL is used to generate the replica interference signal. The AGC in the receiver front-end is used to determine the instant at which the interference signal is present. This interference signal is mixed with the local interference signal (generated by the PLL) to cancel the interference signal. The locally generated interference signal can be differenced from the incoming signal, noise and interference to give signal and noise only, thus cancelling the interference [Cooper and Daly, 1997]. A single PLL can eliminate a single interference signal and hence multiple PLLs are required to eliminate multiple interference signals.

4.4.10 Multilevel Sampling

Leica GPS Inc. developed a technology to mitigate the RFI using a multi-level sampling technique. It employs an adaptive ADC wherein the ADC sampling threshold levels are dynamically controlled by the processor to maintain a statistical frequency which is determined using the quantization thresholds and quantized samples. This technique also improves in-band rejection of the narrowband interference signals [Maenpa et al., 1997]. SAW filters can be used to eliminate the out-of-band interference.

Braasch et al. [1997] analysed an interference suppression unit (ISU), developed by Electro-Radiation Inc., which provides significant interference tolerance. It is effective against different types of interference and can be used with any patch antenna. It was shown to be effective in suppressing an additional 20 dB of broadband noise and narrowband interference compared to stand alone GPS receivers.

4.4.11 Advantage of a Software Receiver

A software receiver allows flexibility in dealing with interference. The exploitation of the spectrum transforms, and other mathematical tools, is more feasible in software than in traditional hardware receivers [Cutright et al. 2003]. Software receivers can look at the signal in different domains and can process a block of data rather than individual samples. This allows for direct filtering of the RFI signals thereby minimizing the effects of RFI. Cutright et al. [2003] developed a frequency domain approach to mitigate the RFI in a software receiver. This method transforms the IF signal into the frequency domain

71

and removes the bias from the spectrum. The bias might be introduced by the receiver front-end bandwidth. The detection threshold is determined to separate the noise and the interference signals. The frequency bins exceeding the threshold are identified and they contain the RFI. These bins are set to zero to remove the interference signal and are transformed back into the time domain. The new signal is free from RFI and is passed to the software correlator. This algorithm is useful for isolating narrow in-band and pulse interference [Cutright et al., 2003]. However, wideband interference cannot be easily separated from the signal.

Burns et al. [2002] evaluated interference mitigation in a software receiver by varying the number of bits in the ADC. The tracking accuracy of the receiver in the presence of interference was chosen as the criteria for determining the success of the interference mitigation. The interference signals were introduced after the receiver started tracking the satellite. The FFT of the incoming data was studied to determine the frequency components with interference. These bins were removed to eliminate the interference signals [Burns et al., 2002]. The signal is better represented with a higher number of the ADC levels and allows for a better estimation of the frequency components containing the interference.

RFI mitigation methods discussed above try to mitigate the interference at various processing stages in the GPS receiver. RFI mitigation at antenna reception eliminates the interference signal before it can enter the receiver. RF filters are essential to eliminate the interference signals outside the GPS frequency band. CW interference can be reduced by preprocessing the IF signal before passing it to the GPS correlator. A swept CW affects all the receiver channels which can be reduced using adaptive notch filters. A blanking method can be used to remove pulse interference. A robust solution to the GPS jamming will always require a variety of anti-jam technologies [Heppe and Ward, 2003].

CHAPTER 5: ACQUISITION: IMPLEMENTATION AND RESULTS

This chapter discusses two acquisition schemes (circular convolution and modified circular convolution) suitable for software receiver implementation. Different acquisition performance parameters are studied using these two schemes.

5.1 Acquisition Implementation

The acquisition process is used to detect the presence of a signal and provide coarse estimates of the code phase and Doppler to the tracking process. It exploits the autocorrelation and cross-correlation properties of the GPS PRN codes to acquire the signal. A block diagram of the acquisition process is shown in Figure 5.1. All the blocks except the acquisition detector and the acquisition manager are common to the tracking process. The acquisition and tracking processes form the core blocks of the correlator in a GPS receiver. Different modules in the acquisition process are discussed below.

Acquisition manager: This module manages the various blocks of the acquisition process and specifies the parameters of operation to each block. It decides the PRN to be searched and the predetection integration time for each cell search. It also specifies the Doppler and code phase range to be searched for the corresponding PRN along with the parameters to compute the detection threshold for acquisition.

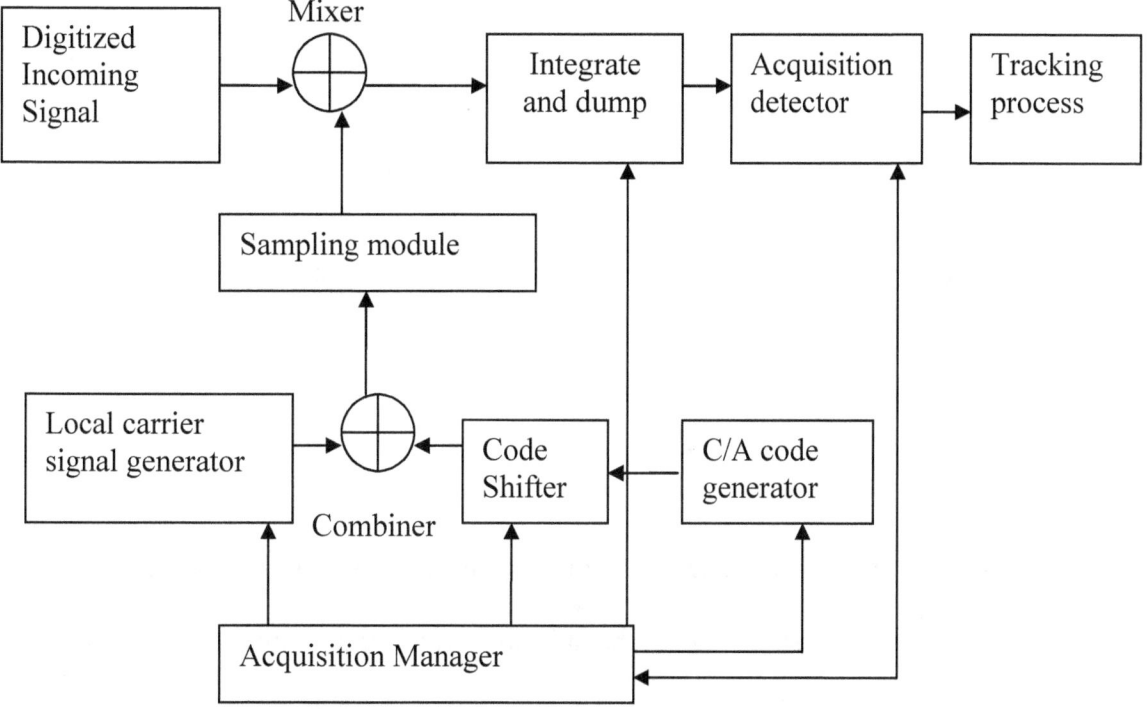

Figure 5.1: Block diagram of the GPS acquisition process

Local carrier signal generator: This module is used to generate the carrier to match the frequency of the incoming IF signal. It generates a carrier signal with frequency as the sum of the receiver IF and the Doppler frequency to be searched. It generates both the in-phase and quadrature components of the carrier signal. The Doppler frequency is modified after all the cells for that particular Doppler are searched with no success.

C/A-code generator: C/A-code generation is explained in Section 2.4. This module generates the C/A-code for the desired PRN number. The C/A-code generator should be capable of generating the code for all GPS satellites.

Code shifter: This module is used to shift the C/A-code by the code phase amount to be searched. The code phase should be properly matched with the incoming signal to acquire it.

Combiner: This module is used to combine the signals applied at its input. The carrier signal is combined with the shifted C/A-code to obtain a local replica of the incoming signal.

Sampling module: The incoming IF signal is sampled at an appropriate sampling frequency chosen to avoid the aliasing effect and to reduce processing power. The sampling signal used to sample the incoming signal must match in phase with the local signal. If there is a phase mismatch, there will be incorrect representation of the local signal with the incoming signal which will yield incorrect results.

Mixer: It mixes the incoming signal with a local replica signal to perform carrier and code wipe off. The resulting signal consists of two components with frequencies as the sum and the difference of the two signals. Correlation is performed during the code wipe off which yields a correlation peak. The acquisition detector determines whether the correlation peak is correct. The high frequency component at the mixer output needs to be eliminated and the low frequency component should be processed to determine if the acquisition is a success.

Integrate and dump: This section integrates the mixer output and acts as a low pass filter (LPF) to eliminate the high frequency component. The integrated signal is combined across the integration periods before passing it to the acquisition detector.

Acquisition detector: This module is used to detect the presence of the GPS signal. Noise computation is an important part of the acquisition process. Detection threshold computation is explained in Section 5.2 and is the minimum noise level which the correlation peak should exceed to be detected as a signal. It should be optimally chosen to avoid a false lock and to allow weak signal acquisition. A signal is acquired when the correlation peak exceeds the detection threshold and estimates of the code phase and Doppler of the cell under search are passed to the tracking process. If a signal is not detected, the acquisition manager searches the next cell. Once all the cells are exhausted the next GPS satellite is searched and the process is repeated.

5.2   Detection Threshold

Acquisition is a two-dimensional (code phase and Doppler) search process whereby the search range is decided by a priori knowledge of the satellite positions. If this information is not available then a sky search has to be performed. The Doppler range should be carefully chosen to include the frequency uncertainty resulting from satellite motion, user dynamics and the receiver clock offset. A Doppler bin is defined as $2/(3T)$, where T is the signal integration time or dwell time per cell in seconds [Tsui and Bao, 2000]. The dwell time should be longer to acquire weak signals. However, the actual signal strength received depends upon the signal environment and is not known until the SV signal is acquired. The SV signal power decreases under foliage condition, urban canyon and indoor environments.

The search pattern is usually in the code phase direction from 1 to 1023 at a constant Doppler bin. In the Doppler search direction, the search pattern is typically from the mean value of the Doppler uncertainty and then symmetrically one Doppler bin at a time on either side of the mean Doppler value until the 3-sigma Doppler uncertainty has been searched [Tsui and Lin, 2001]. The integrate dump module integrates the in-phase (I) and the quadrature (Q) signals over the dwell time for each cell. Then the envelope $\sqrt{I^2 + Q^2}$ is computed and compared to a threshold to determine the presence of the SV signal. The signal detection is a statistical process because each cell contains noise with or without the signal [Kaplan, 1996]. Each case has its own probability density function (PDF) which is shown in Figure 5.2.

The PDF for the noise without the signal has a zero mean while the PDF for the noise with the signal has a non-zero mean. The detection threshold is usually based on an acceptable false detection probability. If the envelope obtained from $\sqrt{I^2 + Q^2}$ is above the detection threshold the signal is present otherwise it is deemed as noise. Figure 5.2 illustrates the four outcomes from a single trial process with two being right and two being wrong. The detection threshold can be properly computed knowing both PDFs of the envelopes and it depends upon the single trial probability of detection (P_d) and single trial probability of false alarm (P_{fd}) defined in Equations (5.1) and (5.2).

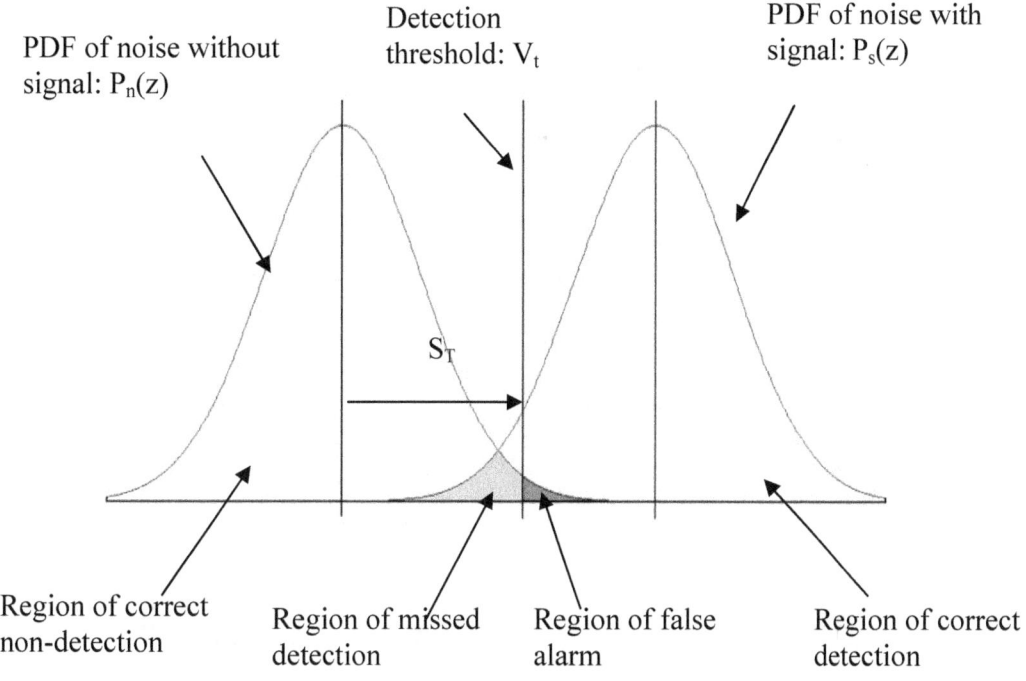

Figure 5.2: PDF of noise and signal

$$P_d = \int_{V_t}^{\infty} P_s dz \qquad\qquad 5.1$$

$$P_{fd} = \int_{V_t}^{\infty} P_n dz \qquad\qquad 5.2$$

where

$P_s(z)$ is the PDF of the envelope in the presence of the signal, and

$P_n(z)$ is the PDF of the envelope in the absence of the signal.

The I and Q signals have a Gaussian distribution which causes the envelope formed by

$\sqrt{I^2 + Q^2}$ to have a Ricean distribution defined in Equation (5.3).

$$P_S(z) = \frac{z}{\sigma_n{}^2} e^{\frac{-(z+A^2)}{2\sigma_n{}^2}} \, I_0\left(\frac{zA}{\sigma_n{}^2}\right) \text{ for } z \geq 0; \ P_S(z) = 0 \text{ otherwise} \qquad 5.3$$

where

 z   is the random variable,

 σ_n is the RMS noise,

 A   is the RMS signal amplitude, and

 $I_0\left(\dfrac{zA}{\sigma_n{}^2}\right)$ is the modified Bessel function of zero order.

Equation (5.3) can be expressed in terms of the predetection SNR (s/n) as given in Equation (5.4).

$$P_S(z) = \frac{z}{\sigma_n{}^2} e^{-\left(\frac{z^2}{2\sigma_n{}^2} + s/n\right)} I_0\left(\frac{z\sqrt{2s/n}}{\sigma_n}\right) \qquad 5.4$$

where

 s/n   is the predetection SNR $= A^2/2\sigma_n{}^2$ (power ratio) $= 10^{(S/N)/10}$,

 S/N   is the predetection SNR in dB $= C/No + 10\log T$ (dB),

 C/No   is the carrier to noise power density ratio in dB, and

 T   is the search dwell time = predetection integration time.

The PDF for the envelope without the signal present can be obtained by setting the signal amplitude (A) to zero in Equation (5.3). This yields a Rayleigh distribution as defined in Equation (5.5).

$$P_n(z) = \frac{z}{\sigma_n^2} e^{-\left(\frac{z^2}{2\sigma_n^2}\right)} \qquad 5.5$$

Equation (5.6) is obtained by integrating the results of substituting Equation (5.5) into Equation (5.2).

$$P_{fd} = e^{-\left(\frac{V_t^2}{2\sigma_n^2}\right)} \qquad 5.6$$

Equation (5.6) is rearranged to obtain an expression for the detection threshold (V_t) in terms of the single trail probability of false alarm and measured 1-sigma noise power as in Equation (5.7) [Kaplan, 1996].

$$V_t = \sigma_n \sqrt{-2 \ln P_{fd}} \qquad 5.7$$

The 1-sigma noise power was found to be insufficient to prevent false locks for the software receiver implemented by the PLAN group. Hence, a 3-sigma noise power along with a standard value of 10% for the false probability detection was used during the analysis.

APPENDIX A: Circular convolution method

The Fourier transform decomposes or separates a waveform or function into sinusoids of different frequencies which sum to the original waveform [Brigham, 1988]. It identifies or distinguishes the different frequency sinusoids and their respective amplitudes. The Fourier transform of a function f(x) is defined in Equation (A.1) [Brigham, 1988].

$$F(s) = \int_{-\infty}^{\infty} f(x)e^{-i2\pi xs}dx \qquad \text{A.1}$$

The inverse Fourier transform of F(s) is given by Equation (A.2) [Brigham, 1988].

$$f(w) = \int_{-\infty}^{\infty} F(s)e^{i2\pi ws}ds \qquad \text{A.2}$$

The functions and transforms occupy the two domains referred to as the function and the transform. However, in most applications these domains are called as the time and frequency domains. The time domain correlation property for the Fourier transform is discussed below. The correlation of two signals f(x) and g(x) is defined in Equation (A.3)

$$h(x) = \int_{-\infty}^{\infty} f(u)g(x+u)du \qquad \text{A.3}$$

Taking the Fourier transform of the correlated signal h(x), Equations (A.4) and (A.5) are obtained.

$$H(s) = \Gamma\{h(x)\} \qquad \text{A.4}$$

$$H(s) = \int_{-\infty}^{\infty} h(x)e^{-i2\pi sx}dx \qquad \text{A.5}$$

82

$$H(s) = \int\limits_{-\infty}^{\infty} [\int\limits_{-\infty}^{\infty} f(u)g(x+u)du]e^{-i2\pi sx}dx \qquad\qquad A.6$$

$$H(s) = \int\limits_{-\infty}^{\infty} f(u)e^{-i2\pi s(-u)}\{\int\limits_{-\infty}^{\infty} g(x+u)e^{-i2\pi s(x+u)}dx\}du \qquad\qquad A.7$$

$$H(s) = \int\limits_{-\infty}^{\infty} f(u)e^{-i2\pi s(-u)}du \; G(s) \qquad\qquad A.8$$

$$H(s) = F^*(s)G(s) \qquad\qquad A.9$$

where

$F^*(s)$ is conjugate of the Fourier transform $F(s)$.

Equation (A.9) is the statement of the correlation theorem of Fourier transform. If $f(x)$ and $g(x)$ are the same function, the integral in Equation (A.3) is normally called the autocorrelation function. If they differ it is called the cross-correlation function. Thus correlation in the time domain is multiplication of the Fourier transform of the two signals. The correlation interval is spread over the entire range of the function. The acquisition search range is two-dimensional in a GPS receiver. The correlation can be reduced to one-domain by using this correlation property of the Fourier transform. The multiplication of the two signal spectrums gives correlation values over the entire range. Thus correlation of the incoming GPS signal with a local signal at a particular Doppler over the entire code phase range is basically multiplication of the Fourier transforms of the incoming GPS signal and the local signal. Thus a search need not be conducted in the code phase domain. This reduces the search domain to a single-dimension (Doppler search) during acquisition. This method is called circular convolution.

REFERENCES

Agilent Technologies (2003), Understanding Dynamic Signal Analysis, Fundamentals of Signal Analysis Series, Agilent Technologies, Application note 1405-2.

Akopian D., H. Valio and S. Turunen (2002), Fine Frequency Resolution Acquisition Methods for GPS Receivers, Proceedings of ION GPS 2002, Portland, OR, September 24-27, pp. 2515-2523.

Akos D., P.L. Normark, A. Hansson, A. Rosenlind and P. Enge (2001), Real-Time GPS Software Radio Receiver, Proceedings of ION NTM 2001, Long Beach, CA, January 22-24, pp. 809-816.

Alaqeeli A. and J. A. Starzyk, (2001), Hardware Implementation of Fast Convolution for GPS Signal Acquisition Using FPGA, Proceedings of 33rd South-eastern Symposium on System Theory, Athens, OH, March 2001, pp 17-204.

Alaqeeli A. (2002), Global Positioning System Signal Acquisition and Tracking Using Field Programmable Gate Arrays, PhD thesis, Ohio University, November 2002.

Bastide F., D. Akos, C. Macabiau and B. Roturier (2003), Automatic Gain Control (AGC) as an Interference Assessment Tool, Proceedings of ION GPS 2003, Portland, OR, September 9-12, pp. 2042-2053.

Behre C., R. Ornedo, G. Rogeness and T. Moore (2002), Satellite Acquisition for a Strike Missile under Jamming and Time Initialization constraints, <u>Proceedings of ION NTM 2002</u>, San Diego, CA, January 28-30, pp.254-264.

Betz J.W. (2000), Effect of Narrowband Interference on GPS Code Tracking Accuracy, <u>Proceedings of ION NTM 2000</u>, Anaheim, CA, January 26-28, pp.16-27.

Bond K. and J. Brading (2000), Location of GPS Interference using Adaptive Antenna Technology, <u>Proceedings of ION GPS 2000</u>, Salt Lake City, UT, September 19-22, pp. 512-518.

Braasch M., C.A. Snyder and R. Olin (1997), Ranging Accuracy Considerations in GPS Interference Suppression, <u>Proceedings of ION GPS 1997</u>, Kansas City, MO, September 16-19, pp. 1483-1487.

Braasch M.S. and F. Van Graas (1991), Guidance Accuracy Considerations for Real time GPS Interferometry, <u>Proceedings of ION GPS 1991</u>, Albuquerque, NM, September 9-13, pp. 373-386.

Brigham E. O. (1988) <u>The Fast Fourier Transform and Its Applications</u>, Englewood Cliffs, Prentice-Hall Inc., New Jersey, NY.

Brown A., D. Reynolds, D. Roberts and S. Serie (1999), Jammer and Interference Location System – Design and Initial Test Results, Proceedings of ION GPS 1999, Nashville, TN, September 14-17, pp. 137-142.

Brown A., S. Atterberg and N. Gerein (2000), Detection and Location of GPS Interference Sources using Digital receiver Electronics, Proceedings of IAIN World Congress and ION AM 2000, San Diego, CA, June 26-28, pp. 269-274.

Buck T. and G. Sellick (1997), GPS RF Interference via a TV Video Signal, Proceedings of ION GPS 1997, Kansas City, MO, September 16-19, pp. 1497-1502.

Burns J., C. Cutright and M. Braasch (2002), Investigation of GPS Software Radio Performance in Combating Narrow Band Interference, Proceedings of ION AM 2002, Albuquerque, NM, June 24-26, pp. 523-530.

Chansarkar M. (2000), Acquisition of GPS Signals at Very Low Signal to Noise Ratio, Proceedings of ION NTM 2000, Anaheim, CA, January 26-28, pp 731-737.

Choi I. H., S. H. Park, D. J. Cho, S. J. Yun, Y. B. Kim and S. J. Lee (2002), A Novel Weak Signal Acquisition Scheme for Assisted GPS, Proceedings of ION GPS 2002, Portland, OR, September 24-27, pp. 177-183.

Cooper J. and P. Daly (1997), Pre-processing of GNSS Signals Subject to Interference, Proceedings of ION GPS 1997, Kansas City, MO, September 16-19, pp. 1437-1446.

Cutright C., J. Burns and M. Braasch (2003), Characterization of Narrow-Band Interference Mitigation Performance Versus Quantization Error in Software Radios, Proceedings of ION AM 2003, Albuquerque, NM, June 23-25, pp. 323-332.

Erlandson R. and R. Frazier (2002), An Updated Assessment of the GNSS L1 Radio Frequency Interference Environment, Proceedings of ION GPS 2002, Portland, OR, September 24-27, pp. 591-599.

Escobar A. and J. Harper (2001), High Temperature Superconducting Filters for GPS Interference Mitigation, Proceedings of ION NTM 2001, Long Beach, CA, January 22-24, pp. 364-368.

FCC Report (2003), Report and Order and Second Further Notice of Proposed Rulemaking, Report No. FCC 03-290, Washington, DC.
http://hraunfoss.fcc.gov/edocs_public/attachmatch/FCC-03-290A1.pdf

Geyer M. and R. Frazier (1999), FAA GPS RFI Mitigation program, Proceedings of ION GPS 1999, Nashville, TN, September 14-17, pp. 107-113.

Gunawardena S. (2000), <u>Feasibility Study For Implementation Of Global Positioning System Processing Techniques In Field Programmable Gate Arrays</u>, Master of Science thesis, Ohio University, November 2000.

Gupta I. (1984), Effect of Jammer Power on the Performance of Adaptive Arrays, <u>IEEE Transaction, Antennas Propagation</u>, vol. AP-32, pp. 933-938.

Gupta I. and T. Moore (2001), Space-Frequency Adaptive Processing for Interference Suppression in GPS Receivers, <u>Proceedings of ION NTM 2001</u>, Long Beach, CA, January 22-24, pp. 377-385.

Gormov K., D. Akos, S. Pullen, P. Enge and B. Parkinson (2000), Generalized Interference Detection and Localization System, <u>Proceedings of ION GPS 2000</u>, Salt Lake City, UT, September 19-22, pp. 447-457.

Hegarty G., A. J. VanDierendonck, D. Bobyn, M. Tran, T. Kim and J. Grabowski (2000), Suppression of Pulsed Interference through Blanking, <u>Proceedings of IAIN World Congress and ION AM 2000</u>, San Diego, CA, June 26-28, pp. 399-408.

Heppe S. and P. Ward (2003), RFI & Jamming and its Effects on GPS Receivers, Based on communication theory, <u>Navtech seminars course 452</u>, Dahlgren, VA, May 5-6.

Herold F. and J. Kaiser (2002), GPS Interference Mitigation, <u>Proceedings of ION AM 2002</u>, Albuquerque, NM, June 24-26, pp. 473-482.

Hopfield, H.S. (1969) Two-Quartic Tropospheric Refractivity Profile for Correcting Satellite Data. <u>Journal of Geophysics</u> Res., 74(18), 4487-4499.

ICD-GPS-200 (2003), <u>Interface Control Document</u>, Navstar GPS Space Segment and Navigation User Interface, ARINC Research Corporation, El Segundo, CA, January 14.

Iltnis R. and G. Hanson (1999), C/A Code Tracking and Acquisition with Interference Rejection using the Extended Kalman Filter, <u>Proceedings of ION NTM 1999</u>, San Diego, CA, January 25-27, pp. 881-889.

Johannessen R., S. Gale and M. Asbury (1990), Potential interference sources to GPS and solutions appropriate for applications to civil aviation, <u>IEEE AES Magazine</u>, vol. 5, Issue 1, January 1990, pp. 3-9.

Johnston H. (1999), A Comparison of CW and Swept CW Effects on a C/A Code GPS Receiver, <u>Proceedings of ION GPS 1997</u>, Kansas City, MO, September 16-19, pp. 149-158.

Kaplan E.D. (1996), <u>Understanding GPS: Principles and Applications</u>, Artech House Inc., Norwood, MA.

Kunysz W. (2001), Advanced Pinwheel Compact Controlled Reception Pattern Antenna designed for Interference and Multipath Mitigation, Proceedings of ION GPS 2001, Salt Lake City, UT, September 11-14, pp. 2030-2036.

Lachapelle G. (2002), Navstar GPS: Theory and Applications, ENGO 625, University of Calgary, Calgary, AB.

Ledvina B.M., S.P.Powell, P.M.Kintner and M.L.Psiaki (2003), A 12-channel Real-Time GPS L1 Software Receiver, Proceedings of ION NTM 2003, Anaheim, CA, pp. 767-782.

Lin D. M., J. B.Y. Tsui, L. Lee, Y. T. Liou and J. Morton (2002), Sensitivity Limit of A Stand-Alone GPS Receiver and An Acquisition Method, Proceedings of ION GPS 2002, Portland, OR, September 24-27, pp. 2515-2523.

Littlepage R. (1999), The Impact of Interference on Civil GPS, Proceedings of ION AM 1999, Cambridge, MA, June 28-30, pp. 821-828.

Macabiau C., O. Julien and E. Chatre (2001), Use of Multicorrelator Techniques for Interference Detection, Proceedings of ION NTM 2001, Long Beach, CA, January 22-24, pp. 353-363.

MacGougan G. (2003) High Sensitivity GPS Performance Analysis in Degraded Signal Environments, UCGE #20176, Department of Geomatics Engineering, University of Calgary, Calgary, AB.

Madhani P. and P. Axelrad (2001), Mitigation of the Near-Far Problem by Successive Interference Cancellation, Proceedings of ION GPS 2001, Salt Lake City, UT, September 11-14, pp. 148-154.

Maenpa J. E., M. Balodis, G. Walter and J. Sandholzer (1997), New Interference Rejection Technology from Leica, Proceedings of ION GPS 1997, Kansas City, MO, September 16-19, pp. 1457-1466.

Manz A., K. Shallberg and P. Shloss (2000), Improving WAAS Receiver Radio Frequency Interference Rejection, Proceedings of ION GPS 2000, Salt Lake City, UT, September 19-22, pp. 471-479.

Moore T. and I. Gupta (2003), The Effect of Interference Power and Bandwidth on Space-Time Adaptive Processing, Proceedings of ION AM 2003, Albuquerque, NM, June 23-25, pp. 337-346.

Ndili A. (1994), GPS Pseudolite Signal Design, Proceedings of ION GPS 1994, Salt Lake City, UT, pp. 1375-1382.

Paddan P., P. Naish and M. Phocas (2003), GPS radio IP design for cellular applications, GPS World, February 2003, pp 30-45.

Parkinson B.W., T. Stansell, R. Beard and K. Gromov (1995), A History of Satellite Navigation, Journal of Institute of Navigation, vol. 42, Special Issue 1, pp. 109-164.

Peterson R., R. Ziemer and D. Borth (1995), Introduction to Spread Spectrum Communications, Prentice Hall Inc., New Jersey, NY.

Pietilä S. and J. Syrjärinne (2000), Improved Method for Satellite Acquisition, Proceedings of ION GPS 2000, Salt Lake City, UT, September19-22, pp. 1957-1961.

Ray J. (2003), Advanced GPS Receiver Technology, ENGO 699.73, University of Calgary, Calgary, AB.

RTCA (2001), Minimum Operational Performance Standards for Global Positioning System/Wide Area Augmentation System Airborne Equipment, Document no. DO-229C, RTCA Inc., Washington, DC, November 28.

Saasamoinen, J. (1971) Atmospheric Correction for the Troposphere and Stratosphere in Radio Ranging of Satellite. International Symposium on the Use of Artificial Satellite, Henriksen (ed.), 3rd Washington, 247-251.

Sastry S. (1997), <u>Analog Modulation Schemes</u>, MIT Media laboratory, Massachusetts Institute of Technology, Cambridge, MA.

<u>http://www.media.mit.edu/physics/pedagogy/fab/fab_2002/help_pages/networking_resources/more%20networking/robotics.eecs.berkeley.edu/_sastry/ee20/modulation/node1.html</u>

Sayed A.H. (2004), <u>Sampling and Aliasing Overview</u>, Digital Signal Processing laboratory, Electrical Engineering department, Univeristy of California at Los Angeles, Los Angeles, CA.

<u>http://www.ee.ucla.edu/~dsplab/sa/over.html</u>

Shau-Shinu J. and P. Enge (2001), Finding Source of Electromagnetic Interference (EMI) to GPS Using Network Sensors, <u>Proceedings of ION NTM 2001</u>, Long Beach, CA, January 22-24, pp. 533-540.

Shashidhar K. (2003), <u>GPS Signal Tap User Guide</u>, Accord Software & Systems Pvt. Ltd., India, <u>www.accord-soft.com</u>.

Spilker J.J. Jr. and B.W. Parkinson (1996), <u>Overview of GPS Operation and Design, Global Positioning System: Theory and Applications</u>, Vol. I, American Institute of Aeronautics and Astronautics Inc., Washington, DC.

Spirent communications (2003), <u>User Manual for the GSS 4765 Interference Simulation System operation with SimGEN for windows</u>, Issue 1.0, Spirent Communications Ltd., pp. 4.1-4.12.

Stenbit J. (2001), <u>Global Positioning System Standard Positioning Service Performance Standard</u>, Command, Control, Communications, and Intelligence, Department of Defense, Washingtion, DC.

Tsui Y. and J. Bao (2000), <u>Fundamentals of Global Positioning System Receivers: A Software Approach</u>, John Wiley & Sons Inc., New York, NY.

Tsui J. B. Y. and D. M. Lin (2001), An Efficient weak signal acquisition algorithm for software GPS receiver, <u>Proceedings of ION GPS 2001</u>, Salt Lake City, UT, September 11-14, pp .115-136.

Vaccaro J. and R. Fante (2000), Ensuring GPS Availability in an Interference Environment, <u>Proceedings of ION GPS 2000</u>, Salt Lake City, UT, September 19-22, pp. 458-461.

VanNee D.J.R. and A.J.R.M. Conen (1991), New Fast GPS code acquisition technique using FFT, <u>IEEE Electronic letters</u>, Vol. 27, No, 2, pp. 158-160.

Ward P. (1996), GPS Receiver Search Techniques, <u>IEEE proceedings of Position Location and Navigation Symposium</u>, Atlanta, GA, April 22-26, pp. 604-611.

Ward P. (1998), Performance Comparisons between FLL, PLL and a Novel FLL-Assisted-PLL Carrier Tracking Loop under RF Interference Conditions, <u>Proceedings of ION GPS 1998</u>, Nashville, TN, September 15-18, pp. 783-795.

Wells D.E., N. Beck, D. Delikaraoglou, A. Kleusberg, E.J. Krakiwsky, G. Lachapelle, R.B. Langley, M. Nakiboglou, K.P. Schwarz, J.M. Tranquilla, and P. Vanicek (1986) <u>Guide to GPS Positioning,</u> Canadian GPS Associates, Fredericton.

Zawistowski T. and P. Shah (2001), <u>An Introduction to Sampling Theory</u>, Department of Electrical and Compute Engineering, University of Houston, Houston, TX.
http://www2.egr.uh.edu/~glover/applets/Sampling/Sampling.html

www.ingramcontent.com/pod-product-compliance
Lightning Source LLC
Chambersburg PA
CBHW080714190526
45169CB00006B/2367